Heinrich Wansing

The Logic of Information Structures

Springer-Verlag

Lecture Notes in Artificial Intelligence 681

Subseries of Lecture Notes in Computer Science
Edited by J. Siekmann

Lecture Notes in Computer Science
Edited by G. Goos and J. Hartmanis

Heinrich Wansing

The Logic of Information Structures

Springer-Verlag
Berlin Heidelberg New York
London Paris Tokyo
Hong Kong Barcelona
Budapest

Series Editor

Jörg Siekmann
University of Saarland
German Research Center for Artificial Intelligence (DFKI)
Stuhlsatzenhausweg 3,
D-66123 Saarbrücken, Germany

Author

Heinrich Wansing
FB Informatik, Universität Hamburg
Bodenstedtstraße 16, D-22765 Hamburg, Germany

CR Subject Classification (1991): I.2

ISBN 3-540-56734-8 Springer-Verlag Berlin Heidelberg New York
ISBN 0-387-56734-8 Springer-Verlag New York Berlin Heidelberg

Printed in Germany

Typesetting: Camera ready by author
45/3140-543210 - Printed on acid-free paper

Preface

The present monograph is a revised version of my doctoral thesis at the Fachbereich Philosophie und Sozialwissenschaften I of the Free University of Berlin. It contains my contribution to the interdisciplinary research project "Systeme der Logik als theoretische Grundlage der Wissens- und Informationsverarbeitung" at the Institute for Philosophy of the Free University of Berlin.

I am very glad that this preface gives me the opportunity to express my gratitude to various people and institutions. First of all, I would like to thank my thesis advisers, David Pearce and Johan van Benthem. With each of them I associate one of the basic ideas which in combination led me to writing this book, viz. (i) to regard negative and positive information as equally relevant and (ii) to vary in a systematic way structural rules of inference. Without their inspiration this thesis would not have come into existence, and it would not have been completed without Johan van Benthem's critical comments and encouragement. Moreover, I gratefully acknowledge scholarships from the Senat von Berlin and the Studienstiftung des deutschen Volkes. To the latter I am indebted for support over many years. A number of colleagues have, in one way or another, contributed to the realization of this book. I would like to thank Kosta Došen and Dirk Roorda for their critical remarks on an early version of Chapter 5 and several stimulating discussions. To Dirk in addition I return thanks for practical help concerning 'huisvesting'. I would also like to thank Peter Schroeder-Heister. His insistance on elimination rules enormously helped me to understand the proof-theoretic approach towards the problem of functional completeness. André Fuhrmann was so kind to comment on an early version of parts of Chapter 9. I also wish to thank Gerd Wagner for numerous conversations and three referees of the *Journal of Logic, Language and Information* for their reports. Finally, I am grateful to Jörg Siekmann for recommending this book for Springer Lecture Notes in AI. From the very beginning I could not have carried out this work without the support from Petra, Kasimir and my parents. They deserve my special, whole-hearted thanks.

As indicated in the text, Chapter 5 is based on 'Formulas-as-types for a hierarchy of sublogics of intuitionistic propositional logic' in D. Pearce & H. Wansing (eds.), *Nonclassical Logics and Information Processing*, Lecture Notes in AI, Vol. 619, Springer-Verlag, Berlin, 1992, and Chapter 4 (the part consisting of Sections 4.1 to 4.4) is (based on) my paper 'Functional completeness for subsystems of intuitionistic propositional logic', forthcoming in the *Journal of Philosophical Logic*. Chapter 4, in particular Section 4.5, is also used in 'On the expressiveness of Categorial Grammar', to appear in J. Wolenski (ed.), *The Legacy of Ajdukiewicz*, Rodopi, Amsterdam. Material from, essentially, Chapter 9 will appear as 'Informational Interpretation of Substructural Propositional Logics' in the *Journal of Logic, Language and Information.*

Hamburg, March 1993 — Heinrich Wansing

Contents

Chapter 1

Introduction

The concept of information can be studied from numerous points of view. Without doubt, however, ***information structure*** and ***information processing*** form central aspects of the study of information. Whereas information structure can be regarded as a subject of ***model theory***, information processing may be viewed as a matter of ***proof theory***. The present investigation pursues this logical perspective. It can be considered a systematic contribution to the line of research that began with S. Kripke's [1965] interpretation of intuitionistic logic in models based on pre-ordered information states. The following table identifies the most important topics that will be dealt with.

proof theory	the systematic variation of structural inference rules in sequent calculi, which offers various options for representing premises as databases and the sequent arrow $\rightarrow$ as an information-processing mechanism (Chapters 2 and 3)
	cut-elimination and consequences thereof (Chapters 3 and 6)
	functional completeness wrt a proof-theoretic interpretation of logical operations (Chapters 4 and 7)
model theory	the encoding of proofs by typed λ-terms and vice versa (Chapters 5 and 8)
	information models, i.e. models based on certain abstract information structures, where by an abstract information structure we understand a non-empty set I viewed as a set of ***information pieces*** or ***information states*** represented by pieces of information together with certain relations or operations on I, possibly some designated pieces of information, and possibly certain conditions on these relations, operations or designated elements (Chapters 1, 2 and 9)

Table 1.1: The main topics.

A general theme, which will be alluded to in considerations on information processing as well as information structure, is the dichotomy between ***positive*** and ***negative*** information (Chapters 2 and 6 - 9). The central claims are that both positive and negative information should be treated in their own right as independent and equally relevant concepts, and that this position leads to strong, *constructive* negation. The first two chapters prepare the stage for a uniform and more comprehensive discussion of information structure and deductive information processing in the remaining chapters by providing examples and motivation.

The whole investigation is concerned with propositional logics only. The propositional sequent calculi considered can easily be extended to predicate logics by adding the usual rules for the existential and universal quantifiers (and, in the presence of strong, constructive negation, the usual rules for their strongly negated forms). We will make use of $\exists$ and $\forall$ as quantifiers in the metalanguage. The metalogic used is classical; repeatedly there will be applications of classical reductio ad absurdum as a rule. Where misunderstandings are unlikely to arise, sometimes no special attention will be paid to the distinction between the mention and use of symbols.

1.1 Intuitionistic propositional logic IPL

An obvious starting point for investigating logics of information structures is reviewing their most famous exponent, viz. intuitionistic propositional logic IPL. Preparatory to the introduction of various formal systems in later chapters, we shall first give a presentation of IPL in perhaps somewhat unorthodox language.

Definition 1.1 The vocabulary of the propositional language L consists of

a denumerable set $PROP$ of propositional variables;
two verum constants: $\mathbf{t}$, $\top$;
one falsum constant: $\bot$;
binary connectives: $/$ (right-searching implication), $\backslash$ (left-searching implication),
$\circ$ (intensional conjunction), $\wedge$ (extensional conjunction), $\vee$ (disjunction);
auxiliary symbols: (,).

Definition 1.2 The set of L-formulas is the smallest set Γ such that

$PROP \subseteq \Gamma$;

$\mathbf{t}, \top, \bot \in \Gamma$;

if $A, B \in \Gamma$, then $(A/B), (A \backslash B), (A \circ B), (A \wedge B), (A \vee B) \in \Gamma$.

We use $p, p_1, p_2, \ldots$ etc. to denote propositional variables, A, B, C, A_1, A_2, ... etc. to denote L-formulas, and X, Y, Z, X_1, X_2, ... etc. to denote finite, possibly empty sequences of L-formula occurrences. Sometimes $<>$ will be used to denote the empty sequence. Outermost parentheses of formulas will not always be written.

Definition 1.3 The notion of a subformula of A is inductively defined as follows:

every L-formula A is a subformula of itself;

the subformulas of A and the subformulas of B are subformulas of (B/A), $(A \setminus B)$, $(A \circ B)$, $(A \wedge B)$, and $(A \vee B)$.

An expression $X \to A$ is called a sequent; X is called its antecedent and A its succedent. In case that $n = 0$, $A_1 \ldots A_n \to A$ denotes $\to A$. Negation in *IPL* is a defined notion, we have $\neg^r A \stackrel{def}{=} (\perp/A)$, $\neg^l A \stackrel{def}{=} (A \setminus \perp)$. $A \rightleftharpoons^+ B$ is used as an abbreviation for $(A \setminus B) \wedge (B \setminus A) \wedge (B/A) \wedge (A/B)$. Next, we present *IPL* as a symmetric sequent calculus with (i) logical rules, (ii) operational rules for introducing connectives on the left hand side (lhs) and on the right hand side (rhs) of the sequent arrow $\to$, and (iii) a number of structural inference rules.

Definition 1.4 The rules constituting *IPL* are:

logical rules :

(id)	$\vdash A \to A$,
(cut)	$Y \to A \quad XAZ \to B \vdash XYZ \to B$,

operational rules :

$(\perp \to)$	$\vdash X\perp Y \to A$,
$(\to \mathbf{t})$	$\vdash X \to \mathbf{t}$,
$(\to \top)$	$\vdash \; \to \top$,
$(\top \to)$	$XY \to A \vdash X\top Y \to A$,
$(\to /)$	$XA \to B \vdash X \to (B/A)$,
$(/ \to)$	$Y \to A \quad XBZ \to C \vdash X(B/A)YZ \to C$,
$(\to \setminus)$	$AX \to B \vdash X \to (A \setminus B)$,
$(\setminus \to)$	$Y \to A \quad XBZ \to C \vdash XY(A \setminus B)Z \to C$,
$(\to \circ)$	$X \to A \quad Y \to B \vdash XY \to (A \circ B)$,
$(\circ \to)$	$XABY \to C \vdash X(A \circ B)Y \to C$,
$(\to \wedge)$	$X \to A \quad X \to B \vdash X \to (A \wedge B)$,
$(\wedge \to)$	$XAY \to C \vdash X(A \wedge B)Y \to C$,
	$XBY \to C \vdash X(A \wedge B)Y \to C$,
$(\to \vee)$	$X \to A \vdash X \to (A \vee B)$,
	$X \to B \vdash X \to (A \vee B)$,
$(\vee \to)$	$XAY \to C \quad XBY \to C \vdash X(A \vee B)Y \to C$;

structural rules :

permutation $(\mathbf{P})$:	$XABY \to C \vdash XBAY \to C$;
contraction $(\mathbf{C})$:	$XAAY \to B \vdash XAY \to B$;
monotonicity $(\mathbf{M})$:	$XY \to B \vdash XAY \to B$.

The rules $(\to /)$ and $(\to \backslash)$ are directional versions of the deduction theorem. The notion of a derivation in IPL of $X \to A$ from a finite, possibly empty sequence of sequent occurrences is defined by induction on the rules of IPL, see Appendix 1.5. If Π is a derivation in IPL of $X \to A$ from the empty sequence, then Π is called a proof of $X \to A$ in IPL, and if there is a proof of $X \to A$ in IPL, this is denoted by $\vdash_{IPL} X \to A$. If the context is clear, we shall sometimes just write $\vdash X \to A$. A proof of $\to A$ in IPL is also called a proof of A in IPL. If there is a proof of A in IPL, then A is called a theorem of IPL. Two formulas A and B are said to be interderivable in IPL, if $\vdash A \to B$ and $\vdash B \to A$, which is abbreviated by $\vdash A \leftrightarrow B$. One can easily show that $\vdash \to A \rightleftharpoons^{+} B$ iff $\vdash A \leftrightarrow B$.

Since **P** is present, directional implications (A/B) and $(B \setminus A)$ resp. directional negations $\neg^{r}A$ and $\neg^{l}A$ are interderivable in IPL. Due to the presence of **M**, the verum constants **t** and $\top$ are interderivable. Moreover, since **C** and **M** are available, also $(A \circ B)$ and $(A \wedge B)$ are interderivable. Thus, in the presence of the structural rules **P**, **C**, and **M** one could do without **t** and intensional conjunction $\circ$, and one could replace the two directional implications $/, \setminus$ resp. negations $\neg^{r}$, $\neg^{l}$ by the more usual implication sign $\supset$ resp. $\neg$.[1] Note that $\top$ (and hence **t**) is definable in IPL as (p/p), for some propositional variable p. In the sequel we shall sometimes use $\supset$ instead of $/, \setminus$ and $\neg$ instead of $\neg^{r}$, $\neg^{l}$, and forget about $\circ$, **t**, and $\top$, if, like in IPL, **P**, **C**, and **M** are assumed to be available. Note also that since **P**, **C**, and the structural inference rule

$$\text{expansion } (\mathbf{E}):\quad XAY \to B \vdash XAAY \to B$$

as a special case of **M** are present, the sequences on the lhs of $\to$ may be conceived of as finite *sets*.

EXAMPLE As an example of a derivation in IPL we prove the distribution of $\wedge$ over $\vee$, i.e. $A \wedge (B \vee C) \to (A \wedge B) \vee (A \wedge C)$, using **M** and **C**:

$$\dfrac{\dfrac{\dfrac{\dfrac{\dfrac{\dfrac{\dfrac{A \to A}{AB \to A} \quad \dfrac{B \to B}{AB \to B}}{AB \to A \wedge B}}{AB \to (A \wedge B) \vee (A \wedge C)} \quad \dfrac{\dfrac{\dfrac{A \to A}{AC \to A} \quad \dfrac{C \to C}{AC \to C}}{AC \to A \wedge C}}{AC \to (A \wedge B) \vee (A \wedge C)}}{A(B \vee C) \to (A \wedge B) \vee (A \wedge C)}}{(A \wedge (B \vee C))(B \vee C) \to (A \wedge B) \vee (A \wedge C)}}{(A \wedge (B \vee C))(A \wedge (B \vee C)) \to (A \wedge B) \vee (A \wedge C)}}{(A \wedge (B \vee C)) \to (A \wedge B) \vee (A \wedge C).}$$

1.2 Kripke's interpretation of *IPL*

We shall briefly describe Kripke's semantics for IPL and reproduce its interpretation in terms of information states as suggested by Kripke.

Definition 1.5 A Kripke frame is a structure $\mathcal{F} = < I, \sqsubseteq >$, where I is a non-empty set and $\sqsubseteq$ is a pre-order (or quasi-order) on I, i.e. $\sqsubseteq$ is a reflexive and transitive binary relation on I.

[1]This is justified on the strength of a replacement theorem that will be proved in Chapter 3.

Definition 1.6 A minimal Kripke model based on a Kripke frame $\mathcal{F}$ is a structure $\mathcal{M} = <\mathcal{F}, v_0>$, where v_0 is a basic valuation function from $PROP \cup \{\bot\}$ into 2^I such that for every $p \in PROP \cup \{\bot\}$ and every $a, b \in I$:

$$\text{if } a \sqsubseteq b, \text{ then } a \in v_0(p) \text{ implies } b \in v_0(p).$$

Definition 1.7 An intuitionistic Kripke model based on a Kripke frame $\mathcal{F}$ is a minimal Kripke model $\mathcal{M} = <\mathcal{F}, v_0>$, where $v_0(\bot) = \emptyset$.

Definition 1.8 Given a Kripke model (minimal or intuitionistic) $\mathcal{M} = <I, \sqsubseteq, v_0>$, v_0 is inductively extended to a valuation function v from the set of all L-formulas into 2^I as follows:

$$\begin{array}{lllll} v(p) & = & v_0(p), & & p \in PROP \cup \{\bot\} \\ v(A \wedge B) & = & v(A \circ B) & = & v(A) \cap v(B), \\ v(A \vee B) & = & v(A) \cup v(B), & & \\ v(A \setminus B) & = & v(B/A) & = & \{a \in I \mid (\forall b \in v(A))\, a \sqsubseteq b \text{ implies } b \in v(B)\}, \\ v(\top) & = & v(\mathbf{t}) & = & I. \end{array}$$

By induction on the complexity of A it can easily be shown that for every Kripke model $\mathcal{M} = <I, \sqsubseteq, v_0>$, every L-formula A, and every a, $b \in I$:

(Heredity) if $a \sqsubseteq b$, then $a \in v(A)$ implies $b \in v(A)$.

Definition 1.9 (semantic consequence) Let $\mathcal{M} = <I, \sqsubseteq, v_0>$ be a Kripke model. A sequent $s = A_1 \ldots A_n \rightarrow A$ holds (or is valid) at $a \in I$

$$\text{iff} \left\{ \begin{array}{ll} a \in v(A_1 \circ \ldots \circ A_n) \text{ implies } a \in v(A) & \text{if } n > 0 \\ a \in v(A) & \text{otherwise.} \end{array} \right.$$

The sequent s holds (or is valid) in $\mathcal{M}$ iff s holds at every $a \in I$. If $\rightarrow A$ holds at $a \in \mathcal{M}$ resp. is valid in $\mathcal{M}$, then also A is said to hold at $a \in \mathcal{M}$ resp. to be valid in $\mathcal{M}$. The sequent s holds (or is valid) in IPL iff s holds in every intuitionistic Kripke model. If $\rightarrow A$ is valid in IPL, then also A is said to be valid in IPL.

If $A_1 \ldots A_n \rightarrow A$ is valid in IPL, then for every intuitionistic Kripke model $\mathcal{M} = <I, \sqsubseteq, v_0>$, $v(A_1 \circ \ldots \circ A_n) \subseteq v(A)$. This notion of validity may be contrasted with the weaker requirement that if $A_1, \ldots, A_n$ are valid in IPL, then A is valid in IPL.

The elements of I can, according to Kripke, be thought of as "points in time (or 'evidential situations'), at which we may have various pieces of information" [1965, p. 98]. We may also identify a state $a \in I$ with the pieces of information available at a. A propositional variable p is verfied at $a \in I$, i.e. $a \in v(p)$, iff there is enough information at a to prove p. Thus, $a \notin v_0(p)$ does not mean that p is falsified at a, it merely says p is not verifed at a. The verification of complex L-formulas at $a \in I$ is determined by the definition of v, given a basic valuation v_0. Since $\neg^r A$ resp. $\neg^l A$ is defined as $\bot/A$ resp. $A \setminus \bot$, for intuitionistic Kripke models we have:

$$\begin{aligned} v(\neg^r A) &= v(\neg^l A) \\ &= \{a \in I \mid (\forall b \in v(A)) \text{ if } a \sqsubseteq b, \text{ then } b \in \emptyset\} \\ &= \{a \in I \mid (\forall b \in I) \text{ if } a \sqsubseteq b, \text{ then } b \notin v(A)\}. \end{aligned}$$

As we are dealing with evidential situations in pre-ordered time, these situations or information states may develop differently depending on the basic information aquired in the course of time. Thus, $a \sqsubseteq b$ says that information state a may develop into information state b, and transitivity of $\sqsubseteq$ becomes rather obvious, intuitively. Moreover, it is assumed that every $b \in I$ may develop into itself, since the information available at b "may be all the knowledge we have for an arbitrarily long time" [Kripke 1965, p. 99]. Eventually, because of (Heredity), information is never lost during the journey through time. Thus, 'possible development' is to be understood as 'possible expansion'.

Theorem 1.10 *IPL* is characterized by the class of all intuitionistic Kripke models, i.e. $\vdash_{IPL} A_1 \ldots A_n \to A$ iff $A_1 \ldots A_n \to A$ is valid in *IPL*.

Soundness, i.e. the 'only if' direction, can be proved by induction on the length of proofs in *IPL*. (Note that the rule M is validity-preserving because $\circ$, which is used to define the evaluation of sequents in Kripke models, is evaluated in exactly the same way as $\wedge$.) Using semantic tableaux, Kripke [1965] shows that every theorem of *IPL* is valid in every intuitionistic Kripke model. We shall sketch a proof of the completeness part of the above theorem (i.e. the 'if' direction) by defining a canonical intuitionistic Kripke model $\mathcal{M}_{IPL} =< I, \sqsubseteq, v_0 >$, i.e. a model which itself characterizes *IPL* (cf. e.g. [Tennant 1978, p. 106 ff.], [Došen 1989, p. 42 f.]). Let Γ be a set of L-formulas; Γ is deductively closed under *IPL* iff $\Gamma = \Gamma \cup \{A \mid\vdash_{IPL} A_1 \ldots A_n \to A \text{ and } A_i \in \Gamma\ (1 \leq i \leq n)\}$. Γ is said to be *IPL*-consistent iff for no sequence $A_1 \ldots A_n$, $A_i \in \Gamma$, $A_1 \ldots A_n \to \bot$ is provable in *IPL*. Γ is called prime iff for all L-formulas A, B: $(A \vee B) \in \Gamma$ implies $A \in \Gamma$ or $B \in \Gamma$. The canonical model $\mathcal{M}_{IPL}$ is defined as follows:

- $I \;=\; \{a \mid a$ is a prime, and IPL–consistent set of L–formulas deductively closed under $IPL\}$,
- $\sqsubseteq$ is the subset relation $\subseteq$,
- $v_0(p) = \{a \in I \mid p \in a\}$,
- $v_0(\bot) = \emptyset$.

Obviously, $\mathcal{M}_{IPL}$ is in fact an intuitionistic Kripke model. It can now be shown that if $\not\vdash_{IPL} A_1 \ldots A_n \to A$, then $A_1, \ldots, A_n$ belong to a prime, *IPL*-consistent set deductively closed under *IPL* which does not contain A. Using this fact, one can prove that for $\mathcal{M}_{IPL} =< I, \sqsubseteq, v_0 >$ the following holds for every L-formula A and every $a \in I$:

(Canon) $a \in v(A)$ iff $A \in a$.

By means of (Canon), completeness can easily be derived. If $A_1 \ldots A_n \to A$ is valid in every intuitionistic Kripke model, in particular it is valid in $\mathcal{M}_{IPL}$. Thus in $\mathcal{M}_{IPL}$, $v(A_1 \circ \ldots \circ A_n) \subseteq v(A)$, if $n > 0$, and $v(A) = I$, otherwise. By (Canon), $\vdash_{IPL} A_1 \circ \ldots \circ A_n \to A$ and thus $\vdash_{IPL} A_1 \ldots A_n \to A$, by $(\to \circ)$ and (cut).

In contrast to the situation in intuitionistic Kripke models, $\perp$ *may* hold at information pieces in minimal Kripke models. As a result of this interpretation of $\perp$, a sequent $X \perp Y \rightarrow A$ is not valid in every minimal Kripke model. The logic characterized by the class of all minimal Kripke models is Johansson's [1937] intuitionistic minimal propositional logic MPL.[2]

Definition 1.11 The rules of MPL are those of IPL without $(\perp \rightarrow)$.

In MPL nothing particular is assumed about $\perp$. The falsum constant $\perp$ can therefore be viewed just as a designated propositional variable used to define intuitionistic minimal negations $\neg^r$, $\neg^l$. The notions of derivation and proof in MPL are defined in the same way as for IPL.

Theorem 1.12 $\vdash_{MPL} A_1 \ldots A_n \rightarrow A$ iff $A_1 \ldots A_n \rightarrow A$ is valid in every minimal Kripke model.

This can be proved in strict analogy to the above proof for IPL. In the canonical model $\mathcal{M}_{MPL}$ for MPL, however, the pieces of information are prime sets of L-formulas deductively closed under MPL which need not be MPL-consistent, and the requirement that $v_0(\perp) = \emptyset$ is dropped.

In the construction of the canonical models $\mathcal{M}_{IPL}$ and $\mathcal{M}_{MPL}$ the defined relation $\sqsubseteq$ is not only a quasi-order but even a partial order on the set of information pieces, i.e. it is also anti-symmetric. Therefore IPL resp. MPL is also characterized by the class of all intuitionistic resp. minimal Kripke models based on a partially ordered set (*poset*). Moreover, there is a standard validity-preserving operation on Kripke models (see e.g. [Kripke 1965]) which applied to an intuitionistic resp. minimal Kripke model based on a *poset* produces an intuitionistic resp. minimal Kripke model $< I, \sqsubseteq, 1, v_0 >$ with $1 \in I$ and where $< I, \sqsubseteq, 1 >$ is a tree (i.e. $< I, \sqsubseteq >$ is a *poset* such that (i) 1 is an initial node: there is no $a \in I$ such that $a \neq 1$ and $a \sqsubseteq 1$, (ii) for every $a, b, c \in I$, if $a \sqsubseteq c$ and $b \sqsubseteq c$, then $a \sqsubseteq b$ or $b \sqsubseteq a$, and (iii) for each $a \in I$, $1 \sqsubseteq_n a$, where $\sqsubseteq_n$ is inductively defined as follows: for every $a, c \in I$, $a \sqsubseteq_0 c$ iff $a = c$; $a \sqsubseteq_1 c$ iff $a \sqsubseteq b$; $a \sqsubseteq_{n+2} b$ iff there is a $c \in I$ such that $a \sqsubseteq_{n+1} b$ and $b \sqsubseteq c$). Thus, IPL resp. MPL is also characterized by the class of all intuitionistic resp. minimal Kripke models based on a tree.[3] Kripke's interpretation can immediately be extended to Kripke models based on a tree: the initial node 1 is to be interpreted as the *initial* piece of information. In Kripke models based on a tree, by (Heredity), the evaluation clause for $\top$ can equivalently be formulated as:

$v(\top) = \{a \mid 1 \sqsubseteq a\}$,

and a sequent $A_1 \ldots A_n \rightarrow A$ can equivalently be said to be valid in a Kripke model $< I, \sqsubseteq, 1, v_0 >$

[2] The implication, negation fragment of MPL was first axiomatized by Kolmogorov [1925].

[3] There is also a standard validity preserving operation converting any Kripke model based on a quasi-ordered set into a Kripke model based on a *poset* (see [Kripke 1965]). Moreover, using a technique which is usually called 'unraveling', any Kripke model based on a tree can be converted into a Kripke model based on a finite tree validating exactly the same L-formulas (see 'selective filtration' in [Gabbay 1981, p. 69 f.]).

$$\text{iff} \begin{cases} 1 \in v(A_1 \circ \ldots \circ A_n) \text{ implies } 1 \in v(A) & \text{if } n > 0 \\ 1 \in v(A) & \text{otherwise.} \end{cases}$$

Thus, a sequent $\rightarrow A$ is provable in IPL resp. MPL iff $\top \rightarrow A$ is provble in IPL resp. MPL iff in every intuitionistic Kripke model resp. minimal Kripke model $< I, \sqsubseteq, 1, v_0 >$, A holds at 1.

The fact that IPL resp. MPL is characterized by the class of all intuitionistic resp. minimal Kripke models based on a tree can be used to show that IPL and MPL enjoy the following form of the disjunction property:

$\Diamond \quad \vdash X \rightarrow A \vee B$ iff $(\vdash X \rightarrow A$ or $\vdash X \rightarrow B)$

(see [van Dalen 1983, p. 186 f.]). We shall now give another proof of (strong) completeness of MPL wrt the class of all minimal Kripke models. For this purpose we will define the canonical model $\mathcal{M}'_{MPL}$ for MPL.

Definition 1.13 The canonical model $\mathcal{M}'_{MPL} =< I, \sqsubseteq, v_0 >$ is defined as follows:

- $I = \{a \mid \exists X = A_1 \ldots A_n (n \geq 0)$ and $a = \{A \mid \vdash_{MPL} X \rightarrow A\}\}$;
- $\sqsubseteq = \subseteq$;
- $v_0(p) = \{a \in I \mid p \in a\}$, for every p in $PROP \cup \{\bot\}$.

It can readily be verified that $\mathcal{M}'_{MPL}$ is in fact a minimal Kripke model. By induction on the complexity of A it can be shown that (Canon) holds for $\mathcal{M}'_{MPL}$. (We use the fact that MPL satisfies $\Diamond$.)

Theorem 1.14 $\vdash_{MPL} A_1 \ldots A_n \rightarrow A$ iff $A_1 \ldots A_n \rightarrow A$ is valid in every minimal Kripke model.

In order to prove completeness, assume that $A_1 \ldots A_n \rightarrow A$ is valid in every minimal Kripke model. Then $A_1 \ldots A_n \rightarrow A$ is valid in $\mathcal{M}'_{MPL}$. Thus, by (Canon), for every $a \in I$, $A_1 \circ \ldots \circ A_n \in a$, implies $A \in a$, if $n > 0$, and $A \in a$, otherwise. By the definition of I, this implies that for every sequence of L-formulas X, if $X \rightarrow A_1 \circ \ldots \circ A_n$ is provable in MPL, then $X \rightarrow A$ is provable in MPL. In particular $A_1 \circ \ldots \circ A_n \rightarrow A$ is provable in MPL.

Note that in $\mathcal{M}'_{MPL}$ every piece of information is finitely represented.

1.3 Grzegorczyk's interpretation of *IPL*

A less well-known semantics for IPL in terms of information pieces has been developed by Grzegorczyk. According to Grzegorkzyk [1964, p. 596] "intuitionistic logic can be understood as the logic of scientific research", where a "scientific research (e.g. an experimental investigation) consists of the successive enrichment of the set of data by new established facts obtained by means of our method of inquiry". In the retrospective,

Grzegorczyk's approach to intuitionistic logic constitutes a concrete version of the characterization of *IPL* by intuitionistic Kripke models based on a tree. Grzegorczyk's approach is *concrete* in the sense that (i) it gives a concrete interpretation to the possible worlds or information pieces instead of taking them as primitive, (ii) for a particular set of information pieces it specifies a particular binary relation on them, and (iii) it specifies a basic valuation function $v_0 : PROP \cup \{\bot\} \longrightarrow 2^I$. In Grzegorczyk's case we have:

- every finite set of propositional variables is a possible world interpreted as a piece of information;
- let I be a nonempty set of information pieces, and let P be a mapping from I in nonempty subsets of I such that

  (∗) if $a = \{p_1, \ldots, p_n\} \in I$, then either $P(a) = \{a\}$ or for every $b \in P(a)$ there exist $p_{n+1}, \ldots, p_{n+k+1}$ $(k \geq 0)$ such that $b = \{p_1, \ldots, p_n, p_{n+1}, \ldots, p_{n+1+k+1}\}$.

  P is interpreted as "the function of possible prolongations of the informations" in I. A binary relation $\sqsubseteq$ on I ("extension of information") is defined in terms of P as follows: for every a, $b \in I$,

  $a \sqsubseteq^0 b$ iff $a = b$;

  $a \sqsubseteq^{n+1} b$ iff there exists a $c \in I$ such that $a \sqsubseteq^n c$ and $b \in P(c)$;

  $a \sqsubseteq b$ iff there exists an $n \in \omega$ such that $a \sqsubseteq^n b$.

Thus, if $a \sqsubseteq b$, then a is a subset of b. A research is defined by Grzegorczyk as a structure $\mathcal{R} =< I, P, 1 >$,[4] where I is a set of information pieces (i.e. a set of finite sets of propositional variables), P is a mapping from I into $2^I - \{\emptyset\}$ satisfying (∗), and every information piece is an extension of the *initial* information piece $1 \in I$: if $a \in I$, then $1 \sqsubseteq a$ (where $\sqsubseteq$ is defined as above). Ideally, $1 = \emptyset$. It can readily be seen that $< I, \sqsubseteq, 1 >$ is a tree. Next, for a given research $\mathcal{R} =< I, P, 1 >$, Grzegorczyk defines a basic valuation function $v_0 : PROP \cup \{\bot\} \longrightarrow 2^I$:

$$v_0(p) = \{a \in I \mid p \in a\}; \quad v_0(\bot) = \emptyset.$$

The basic valuation function v_0 is inductively extended to a valuation function v from the set of all L-formulas into 2^I in exactly the same way as for Kripke models. Thus, every research $< I, P, 1 >$ can be regarded as an intuitionistic Kripke model $< I, \sqsubseteq_P, 1, v_0 >$ based on a tree. Validity of a sequent $s = A_1 \ldots A_n \rightarrow A$ in a research $\mathcal{R} =< I, P, 1 >$ is defined as the validity of s in $< I, \sqsubseteq_P, 1, v_0 >$. Grzegorczyk proves the following characterization theorem:

Theorem 1.15 A is a theorem of *IPL* iff A is valid in every research.

[4] Grzegorcyz uses '0' instead of '1'.

Completeness is proved by Grzegorczyk in an indirect way. He shows that every finite tree T induces a research $\mathcal{R}$ isomorphic to T such that for every L-formula A, A is valid on T according to the topological interpretation of intuitionistic propositional logic[5] iff A is valid in $\mathcal{R}$. Thus, if A is valid in every research $\mathcal{R}$, then it is valid on every finite tree according to the topological interpretation of IPL, and thus it is a theorem of IPL.

Grzegorczyk does not define a canonical research for IPL, and it can easily be shown that such a research doesn't exist. Suppose that $\mathcal{R}$ is a canonical research for IPL with the set of information pieces I. Note that the set $\Gamma = \bigcup\{a \mid a \in I\}$ is finite. Now, take any $q \in PROP$ such that $q \notin \Gamma$. Then $q \setminus p$ is valid in $\mathcal{R}$ for arbitrary p, although $\not\vdash_{IPL} \rightarrow q \setminus p$.[6] Thus $\mathcal{R}$ fails to be canonical.

1.4 The BHK interpretation of IPL

Let us conclude the review of IPL by presenting an interpretation in terms of ***proofs***, viz. the so-called Brouwer-Heyting-Kolmogorov interpretation (BHK interpretation) of the intuitionistic connectives $\wedge$, $\vee$, $\supset$ and the falsum constant $\bot$.[7] To begin with we adopt Girard's [Girard, Lafont & Taylor 1989, p. 5] point of view that "by a ***proof*** we understand not the syntactic formal transcript, but the inherent object of which the written form gives only a shadowy reflection. We take the view that what we *write* as a proof is merely a description of something which is ***already*** a process in itself". From a foundational perspective, the explanatory power of the BHK interpretation depends, of course, on the explanatory power of the notions it makes use of, such as "construction", "transform", etc.[8] In this connection Troelstra and van Dalen [1988, p. 9] point out that "on a very 'classical' interpretation of construction and mapping ... [the interpretation justifies] the principles of two-valued (classical) logic". With these remarks in mind let us consider one recent formulation of the BHK interpretation of IPL.

[Troelstra & van Dalen 1988, p. 9]

(H1) A proof of $A \wedge B$ is given by presenting a proof of A and a proof of B.

(H2) A proof of $A \vee B$ is given by presenting either a proof of A or a proof of B (plus the stipulation that we want to regard the proof presented as evidence for $A \vee B$).

(H3) A proof of $A \supset B$[9] is a construction which permits us to transform any proof of A into a proof of B.

(H4) Absurdity $\bot$ (contradiction) has no proof; a proof of $\neg A$ is a construction which transforms any hypothetical proof of A into a proof of a contradiction.

---

[5] A presentation of the topological semantics for IPL can e.g. be found in [van Dalen 1986].

[6] Grzegorczyk uses '$\supset$' instead of '/' and '\'.

[7] The question of what can be regarded as a proof of a primitive sentence represented by a propositional variable "depends on the particular discipline that is being considered" [López-Escobar 1972, p. 363].

[8] An very eloquent version of the BHK interpretation can be found in [Dragalin 1988, p. 2ff.].

[9] Troelstra and van Dalen use '$\rightarrow$' instead of '$\supset$'.

Generally, the BHK interpretation is regarded as a "natural semantics" [Troelstra & van Dalen 1988, p. 24] for IPL. According to Girard [1989, p. 71] "Heyting's ***semantics of proofs***" even is "[o]ne of the greatest ideas in logic".

1.5 Appendix: Derivations in sequent calculi

Sequent calculi are 'meta-calculi'. A single conclusion sequent calculus acts on sequents $X \to A$, i.e. it manipulates expressions saying that a formula A is a syntactic consequence of a finite sequence of formula occurrences X. At this meta-level we have a syntactic consequence relation $\vdash$ between finite sequences S of sequent occurrences and single sequents. If $\mathcal{L}$ is a logic presented as a sequent calculus, then $\mathcal{D}_{\mathcal{L}}(\Pi, X \to A, S)$, "$\Pi$ is a derivation in $\mathcal{L}$ of $X \to A$ from S" is defined in a way induced by the rules of $\mathcal{L}$. As an example we here give the complete definition for IPL:

- $\mathcal{D}_{IPL}(A \to A, A \to A, <>)$.
- If $\mathcal{D}_{IPL}(\Pi_1, Y \to A, S_1)$ and $\mathcal{D}_{IPL}(\Pi_2, XAZ \to B, S_2)$, then $\mathcal{D}_{IPL}(\frac{\Pi_1\Pi_2}{XYZ \to B}, XYZ \to B, S_1S_2)$.
- $\mathcal{D}_{IPL}(X\bot Y \to A, X\bot Y \to A, <>)$.
- $\mathcal{D}_{IPL}(X \to \mathbf{t}, X \to \mathbf{t}, <>)$.
- $\mathcal{D}_{IPL}(\to \top, \to \top, <>)$.
- If $\mathcal{D}_{IPL}(\Pi, XY \to A, S)$, then $\mathcal{D}_{IPL}(\frac{\Pi}{X\top Y \to A}, X\top Y \to A, S)$.
- If $\mathcal{D}_{IPL}(\Pi, XA \to B, S)$, then $\mathcal{D}_{IPL}(\frac{\Pi}{X \to (B/A)}, X \to (B/A), S)$.
- If $\mathcal{D}_{IPL}(\Pi_1, Y \to A, S_1)$ and $\mathcal{D}_{IPL}(\Pi_2, XBZ \to C, S_2)$, then $\mathcal{D}_{IPL}(\frac{\Pi_1\Pi_2}{X(B/A)YZ \to C}, X(B/A)YZ \to C, S_1S_2)$.
- If $\mathcal{D}_{IPL}(\Pi, AX \to B, S)$, then $\mathcal{D}_{IPL}(\frac{\Pi}{X \to (A\backslash B)}, X \to (A \backslash B), S)$.
- If $\mathcal{D}_{IPL}(\Pi_1, Y \to A, S_1)$ and $\mathcal{D}_{IPL}(\Pi_2, XBZ \to C, S_2)$, then $\mathcal{D}_{IPL}(\frac{\Pi_1\Pi_2}{XY(A\backslash B)Z \to C}, XY(A \backslash B)Z \to C, S_1S_2)$.
- If $\mathcal{D}_{IPL}(\Pi_1, X \to A, S_1)$ and $\mathcal{D}_{IPL}(\Pi_2, Y \to B, S_2)$, then $\mathcal{D}_{IPL}(\frac{\Pi_1\Pi_2}{XY \to (A\circ B)}, XY \to (A \circ B), S_1S_2)$.
- If $\mathcal{D}_{IPL}(\Pi, XABY \to C, S)$, then $\mathcal{D}_{IPL}(\frac{\Pi}{X(A\circ B)Y \to C}, X(A \circ B)Y \to C, S)$.
- If $\mathcal{D}_{IPL}(\Pi_1, X \to A, S_1)$ and $\mathcal{D}_{IPL}(\Pi_2, X \to B, S_2)$, then $\mathcal{D}_{IPL}(\frac{\Pi_1\Pi_2}{X \to (A\wedge B)}, X \to (A \wedge B), S_1S_2)$.
- If $\mathcal{D}_{IPL}(\Pi, XAY \to C, S)$, then $\mathcal{D}_{IPL}(\frac{\Pi}{X(A\wedge B)Y \to C}, X(A \wedge B)Y \to C, S)$.
  If $\mathcal{D}_{IPL}(\Pi, XBY \to C, S)$, then $\mathcal{D}_{IPL}(\frac{\Pi}{X(A\wedge B)Y \to C}, X(A \wedge B)Y \to C, S)$.

- If $\mathcal{D}_{IPL}(\Pi, X \to A, S)$, then $\mathcal{D}_{IPL}(\frac{\Pi}{X \to (A \vee B)}, X \to (A \vee B), S)$.
  If $\mathcal{D}_{IPL}(\Pi, X \to B, S)$, then $\mathcal{D}_{IPL}(\frac{\Pi}{X \to (A \vee B)}, X \to (A \vee B), S)$.
- If $\mathcal{D}_{IPL}(\Pi_1, XAY \to C, S_1)$ and $\mathcal{D}_{IPL}(\Pi_2, XBY \to C, S_2)$, then $\mathcal{D}_{IPL}(\frac{\Pi_1 \Pi_2}{X(A \vee B)Y \to C}, X(A \vee B)Y \to C, S_1 S_2)$.
- If $\mathcal{D}_{IPL}(\Pi, XABY \to C, S)$, then $\mathcal{D}_{IPL}(\frac{\Pi}{XBAY \to C}, XBAY \to C, S)$.
- If $\mathcal{D}_{IPL}(\Pi, XAAY \to B, S)$, then $\mathcal{D}_{IPL}(\frac{\Pi}{XAY \to B}, XAY \to B, S)$.
- If $\mathcal{D}_{IPL}(\Pi, XY \to B, S)$, then $\mathcal{D}_{IPL}(\frac{\Pi}{XAY \to B}, XAY \to B, S)$.

Chapter 2

Generalizations

In this chapter we shall motivate and present certain generalizations of MPL and IPL as logics of information structures. These generalizations point into the direction of (i) systems with a strong, constructive negation $\sim$, i.e. negation in the sense of definite falsity, and (ii) substructural logics,[1] i.e. logics with a restricted set of structural rules of inference. Whereas introducing constructive negation combines virtues of many-valued logic with the intuitionistic implications of MPL and IPL, the variation of structural inference rules offers a number of options for representing deductive information processing. It will be shown that both directions can be entered by a systematic criticism of the BHK interpretation of IPL. Subsequently, various examples of systems are presented which illustrate these richer possibilities.

2.1 Positive and negative information

Any theory of information processing, in order to be viewed as adequate, will be expected to allow for representing both positive as well as negative information. In a reaction to Grzegorczyk's interpretation of IPL, Gurevich [1977] objects that intuitionistic logic does *not* provide an adequate treatment of negative information. In intuitionistic logic (in the language with $\supset$, $\wedge$, $\vee$, and $\bot$) a negated sentence $\neg A$ abbreviates $A \supset \bot$, i.e. $\neg A$ is understood as "A implies absurdity", which is in accordance with Grzegorczyk's intuitions, since he holds that "[t]he compound sentences are not a product of experiment, they arise from reasoning. This concerns also negation: we see that the lemon is yellow, we do not see that it is not blue" [Grzegorczyk 1964, p. 596]. At this point, however, Gurevich observes an unwarranted asymmetry between positive and negative information, since "[i]n many cases the falsehood of a simple scientific sentence can be ascertained as directly (or undirectly) as its truth" [1977, p. 49]. Gurevich therefore would like to have available a primitive *strong* negation in order to express explicit falsity. It is instructive to reformulate Gurevich's point of view in semantical terms. In an intuitionistic Kripke model $< I, \sqsubseteq, v_0 >$, $\neg p$ is true at an information state $a \in I$ iff p is not verified at any information state into which a may develop. Thus, while verifying p at $a \in I$ does not involve considering other information states than a, verifying $\neg p$ involves inspection of all information states $b \in I$ such that $a \sqsubseteq b$. Gurevich's remark

[1] This term has been suggested by Kosta Došen at a conference in Tübingen, October 1990.

amounts to the complaint that there is no possibility of direct falsification of p on the spot. Obviously, Gurevich's insistance on falsification has a famous precursor in Popper's philosophy of science (see e.g. [Popper 1963]) according to which *falsification* is even the more important epistemological principle as compared to *verification*. Now, the idea of taking negative information seriously and putting it on a par with positive information leads Gurevich to intuitionistic logic with strong negation, as developed by Nelson [1949] (and further investigated by Markov [1950]). Recent pleas for the relevance of negative information and the usefulness of strong negation for representing negative reasoning can be found in [Pearce & Wagner 1990], [Pearce 1991], and [Wagner 1991].

Nelson's strong negation $\sim$ is also called *constructive* negation. Indeed, although intuitionistic logic is often referred to as 'constructive logic', intuitionistic negation exhibits certain non-constructive features. Whereas on the one hand, in contrast to classical logic, intuitionistic logic enjoys the disjunction property (or principle of constructible truth):

$(A \vee B)$ is provable iff A is provable or B is provable,

it fails to satisfy the principle of *constructible falsity*, which one should expect to hold for a truly constructive negation:

$\neg(A \wedge B)$ is provable iff $\neg A$ is provable or $\neg B$ is provable.

In Nelson's constructive systems $\mathbf{N}^-$ and $\mathbf{N}$ (see [Almukdad & Nelson 1984], and Section 2.4.1 below)[2] constructible falsity holds wrt $\sim$ (i.e.

$\sim(A \wedge B)$ is provable iff $\sim A$ is provable or $\sim B$ is provable).

The propositional logics $\mathbf{N}^-$ resp. $\mathbf{N}$ result from MPL resp. IPL by conservatively adding $\sim$. In $\mathbf{N}^-$ and $\mathbf{N}$ provable equivalence in the sense of provable mutual implication fails to be a congruence relation, i.e. an equivalence relation respecting the connectives. Therefore intersubstitutivity of provable equivalents fails. Intersubstitutivity holds, however, for formulas A, B provably equivalent in the strong sense that not merely A and B but also their strongly negated forms $\sim A$ and $\sim B$ are interderivable, which reflects the independence and equal importance of positive and negative information. In the standard Kripke semantics for $\mathbf{N}^-$ and $\mathbf{N}$ this is accounted for by distinguishing between truth and falsity conditions. In the Kripke semantics for $\mathbf{N}$ propositional variables may be true, false, or neither true nor false; in the semantics for $\mathbf{N}^-$ they may also be both, true and false.

An unsatisfactory feature of intuitionistic logic besides its having a non-constructive negation is the validity of the principle *ex contradictione sequitur quodlibet*: $(A \wedge \neg A) \supset B$ (or, equivalently, $\perp \supset B$). In terms of information, *ex contradictione* is counterintuitive and should therefore not be accepted. It is not at all clear why contradictory information should make any information whatsoever available, and, of course, usually we are *not* prepared to conclude everything on the basis of maybe just one single contradiction. Thus, a suitable logic of information structures should not only be constructive, it should moreover be similar to MPL and $\mathbf{N}^-$ insofar as it should do without *ex contradictione*. The latter property is usually called paraconsistency (see [Arruda 1980], [Urbas

[2]The propositional logics $\mathbf{N}^-$ resp. $\mathbf{N}$ have independently been developed by von Kutschera [1969], who calls them *direct* resp. *extended direct* propositional logic.

1990]). Assuming paraconsistency doesn't imply that contradictory theories should not be avoided. The point is just that pleading for the avoidance of contradictory theories cannot without further ado be substantiated by pointing out that contradictory information leads to triviality (since it implies everything). Surely, the latter argument is at one's disposal against the background of classical logic; classical logic, however, has already to be rejected as a logic of information structures, because it validates the *tertium non datur* principle: $A \vee \neg A$. And, as Urquhart emphasizes, information may be incomplete: "[w]ith no information whatever about, say, Smith, we can neither conclude "Smith is tall" nor "Smith is not tall". Thus we would *not* expect the law of excluded middle to be valid in a semantics involving pieces of information" [1972, p. 166]. Classical logic also validates $A \supset (B \vee \neg B)$, which is unacceptable, since the information that A just does *not give* us the information that B or the information that $\neg B$. In his discussion of negation in relevance logic, Urquhart describes the problem with negation as the problem of

> [f]inding a semantic evaluation rule ... which is compatible with the existence of both incomplete and inconsistent pieces of information, but otherwise behaves like classical negation [1972, p. 164].

Urquhart then introduces a 'semiclassical' evaluation rule which avoids the validity of $(A \wedge \neg A) \supset B$, $A \supset (B \vee \neg B)$ and *disjunctive syllogism*: $(\neg A \wedge (A \vee B)) \supset B$. *Disjunctive syllogism* should not be accepted from an informational perspective (see also [Dunn 1986, p. 153]): we might have the information that $A \vee B$ just on the strength of the information that A. If now, in addition, the information that $\neg A$ is available, it is simply not justified to conclude that the information that B is available. It may hardly be possible to conceive of 'inconsistent states of affairs' so that in fact, if it is the case that $\neg A$ and it is the case that $A \vee B$, then it is the case that B. Inconsistent information states (or pieces of information), however, should not generally be excluded. Urquhart's semiclassical semantics moreover invalidates contraposition: $(A \supset B) \supset (\neg B \supset \neg A)$ and *reductio ad absurdum*: $(A \supset \neg A) \supset \neg A$, which is all right wrt an informational reading. However, the semiclassical semantics also rules out incomplete information states; it validates *tertium non datur* and is therefore after all to be rejected.

Now, whilst on the one hand (i) classical negation proves unacceptable, because it excludes inconsistent as well as incomplete information states, (ii) Urquhart's semiclassical semantics turns out inappropriate, because it still doesn't allow for incomplete pieces of information, (iii) neither intuitionistic minimal nor intuitionistic negation is constructive, and both invalidate intuitively valid principles like e.g. $A \supset \neg\neg A$, $\neg(A \wedge B) \supset (\neg A \vee \neg B)$, and $\neg(A \supset B) \supset (A \wedge \neg B)$, and (iv) intuitionistic logic validates the *ex contradicitione* principle just as classical logic does, Nelson's system **N** on the other hand (i) unlike intuitionistic minimal and intutitionistic logic is constructive also wrt negation, (ii) invalidates $A \vee \sim A$, $A \supset (B \vee \sim B)$ and allows for incomplete information states, (iii) invalidates $(A \supset B) \supset (\sim B \supset \sim A)$ and $(A \supset \sim A) \supset \sim A$, and (iv) like intuitionistic logic admits of dropping *ex contradictione sequitur quodlibet* (in the form $(A \wedge \sim A) \supset B$) and thereby rendering inconsistent information states possible and invalidating *disjunctive syllogism.* Constructive logic moreover validates the classical De Morgan laws $\sim (A \wedge B) \supset (\sim A \vee \sim B)$, $(\sim A \vee \sim B) \supset \sim (A \wedge B)$,

$\sim (A \vee B) \supset (\sim A \wedge \sim B)$, $(\sim A \wedge \sim B) \supset \sim (A \vee B)$ and the laws of double negation $A \supset \sim\sim A$, $\sim\sim A \supset A$. Thus, constructive negation $\sim$ in Nelson's paraconsistent system $\mathbf{N}^-$ seems to pass Urquhart's informal criteria for a negation in a logic with an interpretation in terms of information pieces.

2.2 The fine-structure of information processing

In deductive information processing the premises are viewed as the database and the consequence relation $\rightarrow$ is taken to be the information-processing mechanism.[3] In a certain sense intuitionistic minimal and intuitionistic logic and also Nelson's systems $\mathbf{N}^-$ and $\mathbf{N}$ constitute *maximal* conceptions of deductive information processing. In all these systems the sequent arrow represents a syntactic consequence relation between finite *sets* of premises and single formulas. These consequence relations are (upwards) monotonic: if formula A is derivable from a finite set Γ of premises, then A is derivable from every finite superset of Γ. If one aims at a formal characterization of our everyday inferences, the monotonicity property, as is well-known, proves to be highly idealized and thus problematic. There are varieties of everyday reasoning which are overtly nonmonotonic: we are deriving conclusions which may turn out wrong in the light of new, additonal information and, accordingly, we are willing to retract, if necessary, such 'provisional' conclusions. Within Artificial Intelligence (AI) nonmonotonicity as a feature of certain kinds of inferences has created a whole field of research with a vast and rapidly growing literature. The most important and well-known AI approaches to nonmonotonic reasoning are perhaps circumscription [McCarthy 1980], default logic [Reiter 1980], autoepistemic logic [Moore 1985], and inheritance networks with exceptions [Horty, Thomason & Touretzky 1987]. In sequent-style presentations of MPL, IPL, $\mathbf{N}^-$ and $\mathbf{N}$ the fact that one is dealing with *monotonic* inferences of single formulas from finite *sets* of premises can explicitly be stated by means of *structural* inference rules, i.e. inference rules which govern the manipulation of premises (or contexts).[4] Besides the monotonicity rule $\mathbf{M}$ there are structural rules allowing for permuting ($\mathbf{P}$) and contracting ($\mathbf{C}$) premise occurrences. If one now considers a systematic variation of such structural rules of inference, the standard package $\{\mathbf{P}, \mathbf{C}, \mathbf{M}\}$ breaks down into a more differentiated ensemble of rules which provides a *fine-tuning* of information processing and the internal structuring of databases (DBs), i.e. DBs need not only be conceived of as sets of sentences with a monotonic inference operation defined on them (for a general

[3] Levesque [1990] calls this the "subjective understanding of logic", i.e. "[r]easoning patterns (or idealizations of them) are modelled by formal derivations in the logic: the rules of inference of the logic are used to model the steps that an agent takes in coming to a conclusion" [1990, p. 266]. This logical representation of reasoning proceeds from the *truth* of the premises to the *truth* of the conclusion; it can be contrasted with what Levesque calls the "objective" use of logic, viz. reasoning from the truth of *belief* in the premises to the truth of *belief* in the conclusion:

$$\vdash A_1 \ldots A_n \rightarrow A \quad \text{versus} \quad \vdash \mathbf{B}A_1 \ldots \mathbf{B}A_n \rightarrow \mathbf{B}A,$$

where $\mathbf{B}$ is a modal belief operator. Exactly the same distinction has also been drawn by Buszkowski [1989] under the labels "external logic" versus "internal logic".

[4] In this respect also (*id*) and (*cut*) are structural rules. Following Girard [Girard, Lafont & Taylor 1989] we will, however, regard (*id*) and (*cut*) as logical rules available in any (ordinary) sequent calculus.

framework of *structured* consequence relations see [Gabbay 1991]).[5]

Once the standard package of structural rules has been called into question, it is obvious to ask what such rules are *natural.* A structural rule which is prominent in AI e.g. is "cautious monotonicity":

$$XY \rightarrow B \quad XY \rightarrow A \vdash XAY \rightarrow B.$$

A more exotic example is provided by the following semi-contraction rule (cf. [Slaney, Surendonk & Girle 1990]):

$$XAABY \rightarrow C \quad XABBY \rightarrow C \vdash XABY \rightarrow C,$$
$$XBAAY \rightarrow C \quad XBBAY \rightarrow C \vdash XBAY \rightarrow C.$$

Being a part of classical logic and non-classical systems like MPL, IPL, $\mathbf{N}^-$, and $\mathbf{N}$, clearly a certain degree of naturalness can be assigned to the usual rules **P**, **C**, and **M**. Therefore in what follows we shall confine ourselves to variations of these structural rules, which are obtained by giving up the assumption of having them available as a package.

Although in the most commonly used logics, **P**, **C**, and **M** are assumed, giving up all or part of them has a long tradition. For example, relevant implicational logic $R_\supset$ developed by Church [1950] and Moh [1950] is nothing but intuitionistic implicational logic $IPL_\supset$ without the monotonicity rule, and in general not to accept the full strength of monotonicity forms the basic idea of relevance logic (cf. [Dunn 1986]). Conceptions of deductive information processing weaker than the intuitionistic minmal one can also typically be found within logical syntax, i.e. Categorial Grammar. The ('product-free' version of the) syntactic calculus of Lambek [1958] e.g. turns out to be intuitionistic implicational logic without any structural rules of inference (but restricted to derivations from non-empty sequences). This syntactic calculus is an order-sensitive logic of *occurrences*, because in syntax every occurrence of a linguistic item to which a syntactic type (logically speaking, a premise) is assigned matters. If the product-free Lambek Calculus is extended by the structural rule of permutation, one obtains the so-called non-directional Lambek Calculus of syntactic categories (see [van Benthem 1986, 1988]). In this case one is concerned with nonmonotonic inferences of single formulas from finite, non-empty *multisets* of formulas, i.e. collections in which every occurrence matters but

[5]In this connection one should also mention Display Logic **DL**, a general and elegant proof-theoretic framework developed by Belnap [1982]. In **DL** there are *formula-connectives* which map formulas into formulas, and moreover there are *structure-connectives* which map structures into structures. The connectives are indexed as belonging to various families of connectives (Boolean, relevance, modal $S4$ etc.). The formula-connectives have a fixed set of rules across all families, and the structure-connectives have a fixed set of basic rules (structural rules) accross all families. These basic structural rules induce a relation called *display equivalence*. They are 'geometrical' rules which allow every 'positive' part of the antecedent (succedent) and 'negative' part of the succedent (antecedent) of a sequent to be displayed as the antecedent (succedent) of a display-equivalent sequent. Within one family, the formula-connectives (and also the structure-connectives) are characterized by assuming certain sets of further structural rules for the structure-connectives in this family. A family may contain formula-connectives (and hence structure-connectives) from various other families. Drawing on Belnap's work, Schroeder-Heister [1991] presents the idea of developing a "universal structural framework" [1991 p. 393] that allows one "to treat all *logical* content of formulas by means of a database of rules" [1991, p. 386].

the order of occurrences is irrelevant. Allowing for derivations from the empty multiset, the non-directional Lambek Calculus turns out to be the implicational fragment of Girard's intuitionistic linear logic without 'exponentials' ([Girard 1987]), i.e. intuitionisitic logic without the rules of monotonicity and contraction (cf. also [Avron 1988], [Troelstra 1992]).

From the point of view of the fine-structure of information processing it becomes clear that the general aim of our investigation cannot be a single formal system, being the one and only logic of information structures. This would be too narrow a perspective, given the fact that possible applications may call for a considerable degree of flexibility. Thus, what we will be looking for instead is a whole *family* of logics which (i) differ wrt to the structural inference rules assumed, and (ii) in particular do justice to negative information as an independent epistemic dimension. Although the constructive minimal family, i.e. the family of substructural subsystems of $\mathbf{N}^-$, will be the preferred framework for deductive information processing, our general attitude and methodology is a more pragmatic one. Therefore intuitinionistic minimal and intuitionistic information processing will also be considered before in later chapters we come to the constructive minimal and the constructive systems.

2.3 Critical remarks on the BHK interpretation of *IPL*

Since its first explicit formulation by Heyting in 1934 (see also [Heyting 1956]) the BHK interpretation has been given a number of slightly different formulations. As we shall see, some clauses in these versions of the BHK interpretation are to a certain extent *ambiguous*. Although we completely agree that "[u]ndeniably, Heyting semantics is very original: it does not interpret the logical operations by themselves, but by abstract constructions" [Girard, Lafont & Taylor 1989, p. 6] or proofs, we will emphasize that the BHK interpretation suffers from a more serious weakness than being ambiguous, viz. its treatment of negation, which is not quite satisfactory from a foundational point of view. We will point out that the BHK interpretation, which is often called proof interpretation, is naturally supplemented by a *disproof* interpretation which leads to a semantical foundation of constructive logic with strong negation rather than intuitionistic logic.

Let us in addition to the version of the BHK interpretation from [Troelstra & van Dalen 1988] presented in the previous chapter consider a few more recent formulations.

[Troelstra 1981, p. 17]

(i) p proves $A \wedge B$ iff p is a pair $< \mathrm{p}_1, \mathrm{p}_2 >$ such that p_1 proves A and p_2 proves B.

(ii) p proves $A \vee B$ if p is either a proof of A or a proof of B.

(iii) p proves $A \supset B$ if p is a construction which transforms any proof q of A into a proof p(q) of B. ...

(iv) p proves $\neg A$ if p proves $A \supset \bot$, that is to say p is a construction which reduces any alleged proof of A to an absurdity.

[van Dalen 1986, p. 231]

(i) a is a proof of $A \wedge B$ [6] iff a is a pair (a_1, a_2) such that a_1 is a proof of A and a_2 is a proof of B.

(ii) a is a proof of $A \vee B$ iff a is a pair (a_1, a_2) such that $a_1 = 0$ and a_2 is a proof of A or $a_1 = 1$ and a_2 is a proof of B.

(iii) a is a proof of $A \supset B$ iff a is a construction that converts each proof b of A into a proof $a(b)$ of B.

(iv) nothing is a proof of $\bot$ (falsity).

[Girard, Lafont & Taylor 1989, pp. 5/6]

2. A proof of $A \wedge B$ is a pair (p, q) consisting of a proof p of A and a proof q of B.

3. A proof of $A \vee B$ is a pair (i, p) with:

- $i = 0$, and p is a proof of A, or
- $i = 1$, and p is a proof of B.

4. A proof of $A \supset B$[7] is a function f, which maps each proof p of A to a proof $f(p)$ of B.

5. In general, the negation $\neg A$ is treated as $A \supset \bot$ where $\bot$ is a sentence with no possible proof.

It has convincingly been argued in [Dummett 1977, chap. 7] and [Prawitz 1977] that instead of just the informal notion of proof as used in the above clauses one should use the informal notion of *direct* or *canonical* proof. In the formal framework of sequent calculi the canonical proofs are, of course, the (*cut*)-free proofs. When in what follows we talk about proofs, we shall (tacitly) mean canonical proofs. Moreover, Dummett [1977] emphasizes that the BHK interpretation is based on the principle of *molecularity* (or compositionality) of interpretation or, more specifically, meaning: the interpretation of complex formulas is explained with reference to the interpretation of their immediate parts. According to Ruitenburg [1991] "the proof interpretation is not reductive: It doesn't express the interpretations of implication ... in simpler terms" [1991, p. 274]. The latter point of view can be compared with saying that the truth conditions for $\supset$ in Kripke models $< I, \sqsubseteq, v >$ for *IPL* are not reductive, since they refer to the quasi-order $\sqsubseteq$. Clearly, non-reductiveness and molecularity do not contradict each other.

According to Kreisel [1965] the clause for implication should be expanded by a correctness postulate saying that there is a proof of the fact that the construction or function referred to in the clause is in fact as required. This extra condition, however, is highly controversial (see e.g. [Prawitz 1977, p. 27], [Sundholm 1983], [Girard, Lafont & Taylor 1989, p. 7], [Girard 1989, p. 71]) and we will refrain from postulating it.

[6] Instead of 'A', 'B' and '$\rightarrow$' van Dalen uses 'φ', 'ψ' and '$\supset$', respectively.

[7] Girard et al. use '$\Rightarrow$' instead of '$\supset$'.

2.3.1 Ambiguity of the BHK interpretation

Let us first consider clause (H1) of [Troelstra & van Dalen 1988], i.e.,

(H1) A proof of $A \wedge B$ is given by presenting a proof of A and a proof of B.

Although at first sight (H1) seems completely clear, it leaves room for a disambiguation:

(DA1) A proof of $A \wedge B$ is given by presenting one proof which proves A as well as B;

(DA2) A proof of $A \wedge B$ is given by presenting a combination of two possibly distinct proofs, one of A and the other of B.

Using (DA1), $\wedge$ can immediately be seen to be commutative, associative, and idempotent: each proof of $A \wedge B$ *is* already a proof of $B \wedge A$ etc. Using (DA2) without making further assumptions, $\wedge$ need not display all of these properties. Clearly, the properties $\wedge$ has under the interpretation (DA2) depend on what is meant by the combination of proofs. According to [Troelstra 1981] (DA2) is the intended reading of (H1), and forming *ordered pairs* is the intended mode of combination of proofs. However, the formation of ordered pairs seems to be inappropriate for intuitionistic conjunction $\wedge$, since the former is neither idempotent, nor associative, nor commutative. (Also pairing in the sense of forming unordered pairs would not do, since a proof Π differs from $\{\Pi\}$ $(= \{\Pi, \Pi\})$ and thus $\wedge$ would fail to be idempotent.) Thus, although on the one hand talking about ordered pairs of proofs is disambiguating (H1), it is on the other hand *by itself* not suitable for explaining intuitionistic conjunction. Now, this can be viewed as requiring simply too much. Indeed, the clauses for $\wedge$ referring to ordered pairs are correct, if a certain notion of valid sequent (or valid consequence)[8] is assumed, which, however, seems not at all to be indissolubly tied up with the above integral clauses for the connectives:

Definition 2.1 A sequent $A_1 \ldots A_n \rightarrow A$ is *valid* iff the following holds:

there exists a construction Π such that $\Pi(< \Pi_1, \ldots, \Pi_n >)$ proves A, whenever $\Pi_1, \ldots, \Pi_n$ prove $A_1, \ldots A_n$, respectively, if $(1 \leq n)$;
there exists a construction that proves A, otherwise.

Thus, $A \wedge B \rightarrow B \wedge A$ e.g. is valid due to the operation of reversing the order of pairs, or, more precisely, due to assuming the existence of this operation. The general problem that comes along with this definition of validity is to lay down and moreover to justify what are admissible constructions. In the clause for $\wedge$ we may e.g. also use unordered pairs, if transitions from Π to $\{\Pi\}$ and $\{\Pi_1, \Pi_2\}$ to $\{\Pi_1\}$, $\{\Pi_2\}$ are regarded as admissible.

Note that instead of assuming the existence of appropriate constructions as required by the definition of validity we could as well use (DA2), assume pairing in the sense of bracketing or juxtaposition as the basic mode of combination (at the level of "shadowy

[8]Explicit statements can e.g. be found in [López-Escobar 1972, p. 367] and [McCarty 1983, p. 124].

reflections"), and specify additional constraints on the combination of proofs: idempotence, associativity and commutativity in the case of bracketing; idempotence and commutativity in the case of juxtaposition.

(DA1) and (DA2) correspond to a distinction which is well known from relevance (and later also from linear) logic, viz. the distinction between 'extensional' ('additive') conjunction $\wedge$ and 'intensional' ('multiplicative') conjunction $\circ$. Let us choose juxtaposition as the basic mode of combining proofs. Then (DA2) can be used to interpret the associative connective $\circ$ syntactically characterized by the sequent rules $(\rightarrow \circ)$ and $(\circ \rightarrow)$. The sequent rules for the extensional connective $\wedge$ are $(\rightarrow \wedge)$ and $(\wedge \rightarrow)$.

Also various clauses given for implication are somewhat ambiguous, but in a less perspicuous way. Consider [van Dalen 1986]. The clause (iii) van Dalen gives for implication requires an understanding of the notions "construction" and "converts". Since these notions are not used as technical terms with a fixed meaning, it might make a difference whether a construction a converts a proof b if a is combined with b (in this order, i.e. ab) or if b is combined with a (i.e. ba). In other words, the explanans leaves room for directionality which is not indicated by the explanandum. There are at least two ways of disambiguating. The first one again points in the direction of substructural logics: distinguish between the two directional implications known from Categorial Grammar (see [Lambek 1958]), the left-searching $\backslash$ (also called 'left residuation') and the right-searching $/$ ('right residuation'), which are syntactically characterized by the sequent rules $(\rightarrow /)$, $(/ \rightarrow)$, $(\rightarrow \backslash)$, and $(\backslash \rightarrow)$. The second method of disambiguation is to work with notions less in need of clarification than "construction" and "convert". Girard's use of the notions "function" and "maps" e.g. seems perfectly all right for the non-directional case when **P** is present.[9]

A logical constant not mentioned in the above versions of the BHK interpretation is the verum constant $\top$. The reason probably is that in intuitionstic logic $\top$ is definable as $p \supset p$, for some propositional variable p. Thus, the identity function is considered as a proof of $\top$. If **M** is dropped from a sequent calculus for IPL, one may distinguish between the two verum constants **t** and $\top$ with sequent rules $(\rightarrow \mathbf{t})$, $(\rightarrow \top)$, and $(\top \rightarrow)$. A BHK-like interpretation of $\top$ and **t** can be given as follows: (i) the empty sequence $<>$ is a proof of $\top$, where clearly for every Π, $<> \Pi = \Pi <> = \Pi$, and (ii) every combination of proofs forms a proof of **t**. Of course, a proof always is a proof of something, and van Dalen [1986, p. 231] explicitly mentions preserving "the feature that from a proof one can read off the result". We shall abbreviate "Π is a proof of A" by $pr(\Pi, A)$. Note that $\top$ and **t** become interderivable if $pr(\Pi_2, A)$ and $pr(\Pi_1\Pi_3, B)$ implies that $pr(\Pi_1\Pi_2\Pi_3, B)$. Also the above mentioned idempotence and commutativity of the combination of proofs are relevant only wrt the results of proofs. Thus, instead of, e.g., idempotence ($\Pi\Pi = \Pi$) it would be enough to require that $pr(\Pi, A)$ implies $pr(\Pi\Pi, A)$ and vice versa.

[9] Note, however, that the very use of the word "function" is criticized by Girard [1989]: "in fact, the interpretation as a function is wrong, since it forgets the dynamics" (p. 83). What Girard suggests is to interpret proofs as *actions* instead of functions (which are usually, although of course not by constructivists, regarded as functional graphs).

2.3.2 The proper treatment of negation

Let us now consider the treatment of negation in the versions of the BHK interpretation in [van Dalen 1986] and [Girard, Lafont & Taylor 1989]. Since $\neg A$ is defined as $A \supset \bot$, a proof of $\neg A$ is a construction resp. function that converts resp. maps each proof of A into a proof of $\bot$. Since there is no (possible) proof of $\bot$, a proof of $\neg A$ would convert any proof of A into a non-existent object. Clearly, it is problematic to explain the meaning of negation by appealing to constructions (functions) that convert (map) objects into non-existent objects, even more so, if we assume that the existence of a proof of $\neg A$ excludes the existence of a proof of A. Troelstra seems to be aware of this fact. He introduces a number of additional notions: we are supposed to know what is a "hypothetical proof" and "a proof of a contradiction" [Troelstra & van Dalen 1988] and what is meant by a reduction of an "alleged proof ... to an absurdity" [1981]. In [Troelstra & van Dalen 1988, p. 9] it is explicitly stated that "[i]n clause H4 the notion of contradiction is to be regarded as a primitive (unexplained) notion". But still, how should something be transformable into a proof of a contradiction? So, if we should add a further primitive notion to the BHK interpretation, absurdity is not a good candidate.

This problem with the BHK interpretation has already clearly been diagnosed by Freudenthal [1937] in his critical remarks on the intuitionistic interpretation of logical formulas:

> Überlegen wir uns darum, was ein negativer Satz aussagt (in diesem Punkte herrscht eine weitgehende Verwirrung: man findet die Behauptung, ein negativer Satz ziele ab auf die Konstruktion eines Widerspruchs; dabei ist völlig unklar, wie irgendwelche Dinge, die man wirklich hergestellt hat, einen Widerspruch enthalten können, überhaupt was *Widerspruch* hier bedeuten soll). Ein negativer Satz $2 \neq 3$ bedeutet, daß in keiner Weise je eine eineindeutige Abbildung der Menge 2 in die Menge 3 gelingen kann; zum Beweis führt man alle (neun) Abbildungen der Menge 2 in die Menge 3 aus und überzeugt sich bei jeder einzelnen davon, daß sie nicht zum Ziele führt. Dies Beispiel enthält vollständig den Mechanismus der negativen Sätze; .... Ein negativer Satz sagt also, daß alle Konstruktionsversuche mit einer bestimmten Zielsetzung scheitern. [1937, p. 113][10]

According to Freudenthal a direct proof of a sentence $\neg A$ would be a demonstration showing that *all* atttempts to prove A fail. As McCarty [1983] shows, Freudenthal's point of view still is very similar to the intuitionistic one. Translated into our notation, McCarty's clause for negation (p. 124) reads as follows:

(iii) $pr(\Pi, \neg A)$ iff for all Π^*, if $pr(\Pi^*, A)$, then $pr(\Pi(\Pi^*), \overline{0} = \overline{1})$,

[10] "Let us therefore reflect upon what a negative sentence expresses (on this point there prevails a large confusion: one can find the claim that a negative sentence aims at the construction of a contradiction; yet it is completely unclear how something which has in fact been constructed may contain a contradiction and, in general, what *contradiction* is intended to mean in this case). A negative sentence $2 \neq 3$ means that in no way ever a $1-1$ mapping from the set 2 into the set 3 can succeed; to prove this one carries out all (nine) mappings from the set 2 into the set 3 and convinces oneself that in each case they do not succeed. This example entirely contains the mechanism of the negative sentences; .... A negative sentence thus says that all attempted constructions with a certain aim fail." (translation HW)

assuming that "no construction proves $\bar{0} = \bar{1}$" (p. 125). Since $\bar{0} = \bar{1}$ has no proof, McCarty replaces (iii) by:

$pr(\Pi, \neg A)$ iff for all Π^*, it is not the case that $pr(\Pi^*, A)$.

This clause obviously is very close to Freudenthal's position; unfortunately, it is blatantly non-constructive, since negation $\neg$ is not interpreted by an abstract construction but by the absence of constructions: if no construction proves A, then any construction proves $\neg A$. It might be objected that the predicate $pr(\, , \,)$ should be decidable. According to Sundholm [1983] it was this requirement which led Kreisel to introduce the so-called "second clause" for implication. We have already mentioned that this second clause is widely regarded as highly problematic. Moreover, even if the $pr(\, , \,)$ predicate is decidable, McCarty's interpretation seems not to be in the spirit of Heyting semantics.

Very similar critical remarks as above have also been put forward by López-Escobar [1972], who suggested to add as a new primitive notion to the BHK interpretation the notion of ***refutation*** (or ***disproof***). As López-Escobar observes, this leads to an interpretation not of intuitionistic logic but rather of Nelson's logic with strong, ***constructive*** negation $\sim$ without *ex contradictione quodlibet*, i.e. Nelson's constructive system $\mathbf{N}^-$. Similarly, considering refutability as the mirror-image of provability, von Kutschera [1969] comes to investigate functional completeness of (the propositional part of) $\mathbf{N}^-$, which he calls ***direct*** propositional logic (see Chapter 7). Here is the disproof interpretation of the intuitionistic connectives $\wedge$, $\vee$, and $\supset$ and the constructive negation $\sim$ as presented by López-Escobar:

i.) the construction c refutes $A \wedge B$ [11] iff c is of the form $< i, d >$ with i either 0 or 1 and if $i = 0$, then d refutes A and if $i = 1$ then d refutes B,

ii.) the construction c refutes $A \vee B$ iff c is of the form $< d, e >$ and d refutes A and e refutes B,

iii.) the construction c refutes $A \supset B$ iff c is of the form $< d, e >$ and d proves A and e refutes B, ...

viii.) [t]he construction c refutes $\sim A$ iff c proves A.

A proof of $\sim A$ is then considered as a refutation of A (and not as a proof of $A \supset \bot$). A fundamental assumption made by López-Escobar is that

$\{A \mid \exists \Pi, \, pr(\Pi, A) \text{ and } pr(\Pi, \sim A)\} = \emptyset$.

If we endorse the stronger assumption that

$\{A \mid \exists \Pi_1 \exists \Pi_2, \, pr(\Pi_1, A) \text{ and } pr(\Pi_2, \sim A)\} = \emptyset$,

then the rule

(ex contradictione) $\vdash X(A \wedge \sim A)Y \rightarrow B$

becomes validity preserving.

[11] López-Escobar uses '&' instead of '$\wedge$' and '$-$' instead of '$\sim$'.

In conclusion we may say that supplementing the BHK interpretation by the notion of disproof avoids the above-described problems caused by the non-constructive nature of intuitionistic negation. Moreover, it also is straightforward and intuitively convincing insofar as it reflects a balance between positive and negative information. The ambiguity of the BHK interpretation detected in the previous subsection need not exclusively be viewed as a shortcoming of this approach. On the contrary, the ambiguity will even turn out to be instructive insofar as it reveals parameters which can be modified so as to obtain an unequivocal *semantical framework* rather than one particular interpretation. In Chapter 6 we shall introduce a certain *proof/disproof interpretation* as such a semantical framework and show that this interpretation is sound wrt a broad range of constructive substructural propositional logics, if suitable constraints are imposed on the combination of proofs and disproofs.

2.4 Examples

We present some examples of important constructive and substructural logics and, in particular, their interpretation in terms of information pieces.

2.4.1 Nelson's constructive systems $\mathbf{N}^-$ and $\mathbf{N}$

In a symmetric sequent calculus for the propositional logic $\mathbf{N}^-$ in the language with $\supset$, $\wedge$, $\vee$, $\bot$, and $\sim$, we have in addition to rules for introducing the connectives into premises and conclusions also rules for introducing on the rhs and on the lhs of $\rightarrow$ strongly negated formulas with $\supset$, $\wedge$, $\vee$, or $\sim$ as the main connective (cf. [Almukdad & Nelson 1984] or [Kutschera 1969], [López-Escobar 1972], [Routley 1974], [Akama 1988a][12]). In $\mathbf{N}$ we have in addition to the rule (ex contradictione) also rules for $\bot$ and $\sim\bot$.

Definition 2.2 The rules constituting $\mathbf{N}^-$ are:

(id); (cut);

$<\rightarrow\wedge>$ $\quad X \rightarrow A \;\; Y \rightarrow B \vdash XY \rightarrow (A \wedge B)$;

$<\wedge\rightarrow>$ $\quad XA \rightarrow C \vdash X(A \wedge B) \rightarrow C$,

$\quad XB \rightarrow C \vdash X(A \wedge B) \rightarrow C$;

$<\rightarrow\vee>$ $\quad X \rightarrow A \vdash X \rightarrow (A \vee B)$,

$\quad X \rightarrow B \vdash X \rightarrow (A \vee B)$;

$<\vee\rightarrow>$ $\quad XA \rightarrow C \;\; YB \rightarrow C \vdash XY(A \vee B) \rightarrow C$;

$<\rightarrow\supset>$ $\quad XA \rightarrow B \vdash X \rightarrow (A \supset B)$;

$<\supset\rightarrow>$ $\quad Y \rightarrow A \;\; XB \rightarrow C \vdash X(A \supset B)Y \rightarrow C$;

$<\rightarrow\sim\sim>$ $\quad X \rightarrow A \vdash X \rightarrow\sim\sim A$;

$<\sim\sim\rightarrow>$ $\quad XA \rightarrow B \vdash X \sim\sim A \rightarrow B$;

$<\rightarrow\sim\wedge>$ $\quad X \rightarrow\sim A \vdash X \rightarrow\sim (A \wedge B)$,

$\quad X \rightarrow\sim B \vdash X \rightarrow\sim (A \wedge B)$;

$<\sim\wedge\rightarrow>$ $\quad X \sim A \rightarrow C \;\; Y \sim B \rightarrow C \vdash XY \sim (A \wedge B) \rightarrow C$;

[12] [Akama 1988a] should be read in combination with Tanaka's [1991] critical comments.

$$
\begin{array}{ll}
\langle \rightarrow\sim\vee \rangle & X \rightarrow\sim A \quad Y \rightarrow\sim B \vdash XY \rightarrow\sim (A \vee B); \\
\langle \sim\vee\rightarrow \rangle & X \sim A \rightarrow C \vdash X \sim (A \vee B) \rightarrow C, \\
 & X \sim B \rightarrow C \vdash X \sim (A \vee B) \rightarrow C; \\
\langle \rightarrow\sim\supset \rangle & X \rightarrow A \quad Y \rightarrow\sim B \vdash XY \rightarrow\sim (A \supset B); \\
\langle \sim\supset\rightarrow \rangle & XA \rightarrow C \vdash X \sim (A \supset B) \rightarrow C; \\
 & X \sim B \rightarrow C \vdash X \sim (A \supset B) \rightarrow C;^{13}
\end{array}
$$

P; **C**; **M**.

Definition 2.3 The rules of **N** are those of $\mathbf{N}^-$ plus (ex contradictione), $(\bot \rightarrow)$ and

$$(\rightarrow\sim\bot) \quad \vdash X \rightarrow\sim \bot.^{14}$$

Models for $\mathbf{N}^-$ and **N** can be based on Kripke frames, i.e. pre-orders, $< I, \sqsubseteq >$ (cf. e.g. [Thomason 1969], [López-Escobar 1972], [Routley 1974], or [Gurevich 1977]).

Definition 2.4 (i) A Kripke model for $\mathbf{N}^-$ is a structure $\mathcal{M} = < \mathcal{F}, v_0^+, v_0^- >$, where $\mathcal{F}$ is a Kripke frame and v_0^+, v_0^- are basic valuation functions from $PROP \cup \{\bot\}$ into 2^I such that for every $p \in PROP \cup \{\bot\}$ and every $a, b \in I$:

if $a \sqsubseteq b$, then $a \in v_0^+(p)$ implies $b \in v_0^+(p)$,
if $a \sqsubseteq b$, then $a \in v_0^-(p)$ implies $b \in v_0^-(p)$.

(ii) A Kripke model for **N** is a model for $\mathbf{N}^-$ $< I, \sqsubseteq, v_0^+, v_0^- >$, where for every $p \in PROP$, $v_0^+(p) \cap v_0^-(p) = \emptyset$ and $v_0^+(\bot) = \emptyset$, $v_0^-(\bot) = I$.

Definition 2.5 The valuation functions v^+, v^- induced by a Kripke model for $\mathbf{N}^-$ $< I, \sqsubseteq, v_0^+, v_0^- >$ are the functions from the set of all formulas in $\{\supset, \wedge, \vee, \bot, \sim\}$ into 2^I which are inductively defined as follows (where $p \in PROP \cup \{\bot\}$):

$$
\begin{array}{lcl}
v^+(p) & = & v_0^+(p), \\
v^-(p) & = & v_0^-(p), \\
 & & \\
v^+(A \supset B) & = & \{a \mid (\forall b \in v^+(A)) a \sqsubseteq b \text{ iplies } b \in v^+(B)\}, \\
v^-(A \supset B) & = & v^+(A) \cap v^-(B), \\
 & & \\
v^+(A \wedge B) & = & v^+(A) \cap v^+(B), \\
v^-(A \wedge B) & = & v^-(A) \cup v^-(B), \\
 & & \\
v^+(A \vee B) & = & v^+(A) \cup v^-(B), \\
v^-(A \vee B) & = & v^-(A) \cap v^-(B), \\
 & & \\
v^+(\sim A) & = & v^-(A), \\
v^-(\sim A) & = & v^+(A).^{15}
\end{array}
$$

[13] Note that the formulation of the operational rules takes advantage of the presence of **P**, **C**, and **M**.

[14] Almukdad and Nelson [1984] do not assume $\bot$ to be in the language. Clearly, $\bot$ is definable in **N** as $p \wedge \sim p$, for some propositional variable p. Van Dalen [1986] includes $\bot$ in his presentation of **N**, however, probably because he thinks of **N** as an extension of intuitionistic logic.

By simultanous induction on the complexity of A one can show that for every Kripke model $\mathcal{M} = < I, \sqsubseteq, v_0^+, v_0^- >$ for $\mathbf{N}^-$, every formula A, and every a, $b \in I$,

(Heredity $^+$) if $a \leq b$, then ($a \in v^+(A)$ implies $b \in v^+(A)$),

(Heredity $^-$) if $a \leq b$, then ($a \in v^-(A)$ implies $b \in v^-(A)$),

and moreover, if $\mathcal{M}$ is a Kripke model for $\mathbf{N}$, $v^+(A) \cap v^-(A) = \emptyset$.

REMARK Obviously, instead of the two valuations v^+ and v^-, one could in the case of Kripke models for $\mathbf{N}$ use one three-valued resp. in the case of Kripke models for $\mathbf{N}^-$ one four-valued valuation $\mathbf{v}$ assigning to each pair $< a, A >$ ($a \in I$, A a formula) one of the values t (true), f (false), or u (undetermined), resp. t, f, u, or o (overdetermined):

$$\begin{array}{lll} \mathbf{v}(a,A) = t & \text{iff} & a \in v^+(A), \\ \mathbf{v}(a,A) = f & \text{iff} & a \in v^-(A), \\ \mathbf{v}(a,A) = u & \text{iff} & a \in I - (v^+(A) \cup v^-(A)), \\ \mathbf{v}(a,A) = o & \text{iff} & a \in v^+(A) \cap v^-(A). \end{array}$$

Definition 2.6 (semantic consequence) Let $\mathcal{M} =< I, \sqsubseteq, v_0^+, v_0^- >$ be a Kripke model for $\mathbf{N}^-$. A sequent $A_1 \dots A_n \rightarrow A$ holds (or is valid) in $\mathcal{M}$

$$\text{iff} \begin{cases} v^+(A_1 \wedge \ldots \wedge A_n) \subseteq v^+(A) & \text{if } X \text{ is nonempty,} \\ v^+(A) = I & \text{otherwise.} \end{cases}$$

Kripke's interpretation of IPL in terms of information states or pieces can directly be applied to the above semantics for $\mathbf{N}^-$ and $\mathbf{N}$, except that now truth conditions as specified by valuations v^+ are accompanied by *falsity conditions* as specified by valuations v^-. In contrast to Kripke models for $\mathbf{N}$, Kripke models for $\mathbf{N}^-$ allow for inconsistent pieces of information.

The standard completeness proofs for MPL and IPL can easily be adapted to prove

Theorem 2.7 $\mathbf{N}^-$ resp. $\mathbf{N}$ is characterized by the class of all Kripke models for $\mathbf{N}^-$ resp. $\mathbf{N}$.

2.4.2 Relevant implicational logic $R_\supset$

Relevant implicational logic $R_\supset$ codifies the idea of relevant inference and forms the integral part of relevance logic (see e.g. [Dunn 1986]).

Definition 2.8 The rules constituting $R_\supset$ are $(id), (cut), (\rightarrow /), (/ \rightarrow), (\rightarrow \backslash)$, $(\backslash \rightarrow), \mathbf{P}$, and $\mathbf{C}$.

[15]These truth and falsity conditions with the exception of the clauses for $\supset$ are the core of ***partial logic***, cf. e.g. [Fenstad, Halvorsen, Langholm & van Benthem 1987], [Thijsse 1990]. As extensions of this core, $\mathbf{N}^-$ and $\mathbf{N}$ may be regarded as systems of partial logic (cf. [Blamey 1986, p. 25 f.]). Note that implications ($A \supset B$) are falsified on the spot. This is a significant difference to Veltman's [1981] ***data semantics*** which assigns weaker, 'dynamic' falsity conditions to ($A \supset B$), viz., in our notation, $v^-(A \supset B) = \{a \mid (\exists b \in v^+(A)) a \sqsubseteq b, b \in v^+(A) \text{ and } b \in v^-(B)\}$.

Urquhart [1972] has developed an informational interpretation of $R_{\supset}$. Instead of furnishing a set of information pieces I with a binary relation as in Kripke's semantics for IPL, Urquhart in his semantics for $R_{\supset}$ assumes a binary operation $\oplus$ on I which is to be thought of as the 'addition' of information pieces. Like Grzegorczyk, he postulates an empty piece of information, 1, which is now used to define validity.[16] A frame $\mathcal{F} = < I, \oplus, 1 >$ for $R_{\supset}$ is required to be a semilattice wrt $\oplus$ and with 1 as least element. In other words, the following equations hold for every $a, b, c, \in I$:

$$a \oplus a = a, \quad (a \oplus b) \oplus c = a \oplus (b \oplus c), \quad a \oplus b = b \oplus a, \quad a \oplus 1 = a.$$

A model $\mathcal{M}$ for $R_{\supset}$ is a frame $< I, \oplus, 1 >$ for $R_{\supset}$ together with a basic valuation function v_0 from $PROP$ into 2^I such that $v_0(p) = \{a \in I \mid a \text{ determines } p\}$ or, equivalently (as suggested by an example in [Urquhart 1972]),

$$v_0(p) = \{a \in I \mid p \in a\}.$$

The basic valuation v_0 is then extended to a mapping v from the set of all implicational formulas into 2^I by stipulating:

$$v(A \setminus B) = v(B/A) = \{a \in I \mid (\forall b \in I) \text{ if } b \in v(A), \text{ then } a \oplus b \in v(B)\}.$$

An implicational formula A is said to be valid in a model $\mathcal{M} = < I, \oplus, 1, v_0 >$ for $R_{\supset}$ iff $1 \in v(A)$.

Theorem 2.9 (Urquhart) A is a theorem of $R_{\supset}$ iff A is valid in every model for $R_{\supset}$.

Soundness is proved by induction on the length of proofs of $\rightarrow A$ in $R_{\supset}$. Completeness is shown by defining a simple canonical model $\mathcal{M}_{R_{\supset}} =< I, \oplus, 1, v_0 >$ for $R_{\supset}$, where I is the family of all finite sets of implicational formulas, $\oplus$ is set union, and $1 = \emptyset$. A valuation function v which can be shown to satisfy the requisite conditions is defined by: $a \in v(A)$ iff $a \rightarrow A$ is provable in $R_{\supset}$. If now A is not a theorem of $R_{\supset}$, then $\emptyset \notin v(A)$, thus $\rightarrow A$ is not provable in $R_{\supset}$.

By a suitable definition of semantic consequence, Urquhart's result can easily be extended to a strong characterization theorem.

Definition 2.10 A sequent $A_1 \ldots A_n \rightarrow A$ is valid in a model $< I, \oplus, 1, v_0 >$ for $R_{\supset}$

$$\text{iff} \begin{cases} (\forall a_1 \in I \ldots \forall a_n \in I) a_i \in v(A_i) \text{ implies } a_1 \oplus \ldots \oplus a_n \in v(A) & \text{if } n > 0 \\ 1 \in v(A) & \text{otherwise.} \end{cases}$$

If now $\nvdash_{R_{\supset}} A_1 \ldots A_n \rightarrow A$, then there are $a_1, \ldots, a_n \in I$ such that $\vdash_{R_{\supset}} a_i \rightarrow A_i$ and $\nvdash_{R_{\supset}} a_1 \oplus \ldots \oplus a_n \rightarrow A$. By the definition of v, $A_1 \ldots A_n \rightarrow A$ is not valid in $\mathcal{M}_{R_{\supset}}$. Note that $\mathcal{M}_{R_{\supset}}$ is a free semilattice; $R_{\supset}$ is therefore also characterized by the class of all free semilattices with least element.

In the terminology of [Došen 1990] a valuation v is called multiplicative iff for every $a, b \in I$ and every implicational formula A:

[16] To be precise, Urquhart uses '$\cup$' instead of '$\oplus$' and '0' instead of '1'.

(mult) if $a \in v(A)$, then $a \oplus b \in v(A)$.

As Urquhart observes, the logic characterized by the class of all multiplicative models $< I, \oplus, 1, v_0 >$ is intutionistic implicational logic $IPL_{\supset}$. Given (mult), we have $1 \in v(A)$ iff for every $b \in I$, $b \in v(A)$. Došen [1990] points out that for semilattices it is enough to assume (mult) for propositional variables and then to establish (mult) for every implicational formula A.

In presenting his semilattice semantics, Urquhart maintains that the addition of information pieces "must obviously fulfill the laws of set union" ([1972, p. 160]), i.e. associativity, commutativity and idempotence. When it comes to generalizations of his semantics, however, he also considers relaxing these properties and correlates them to structural rules of inference in Gentzen-style sequent calculi. He mentions e.g. that in the absence of structural rules $\oplus$ is required only to be associative: "this means that, considering the pieces of information ... as listed on sheets of paper, we do not consider two pieces of information to be identical unless they are given in the form of identical lists" [1972, p. 168].

2.4.3 Categorial logics

If **C** is taken away from the rules of $R_{\supset}$ and sequents are restricted to have non-empty sequences on the lhs of the sequent arrow, the resulting system is the product-free, directional Lambek Calculus of syntactic categories, **LP**. Van Benthem [1986, 1988] characterizes **LP** by means of the following canonical model $M_{\mathbf{LP}} =< I, \oplus, v_0 >$ for **LP**, where I is the family of all finite, non-empty multisets of implicational formulas, $\oplus$ is multiset union (which is associative and commutative but not idempotent), v_0 is defined by $a \in v_0(p)$ iff $\vdash_{\mathbf{LP}} a \rightarrow p$, the validity of a sequent $A_1 \ldots A_n \rightarrow A$ is defined as above, and Urquhart's evaluation clause for implications is used. It can then be shown that for every implicational formula A and every $a \in I$, $a \in v(A)$ iff $\vdash_{\mathbf{LP}} a \rightarrow A$. If **P** is dropped from **LP**, one obtains the product-free Lambek Calculus **L**. In the canonical model $\mathcal{M}_{\mathbf{L}} =< I, \oplus, v_0 >$ for **L**, I is the family of all finite, non-empty sequences of occurrences of implicational formulas, $\oplus$ is the operation of juxtaposition (which is associative but neither commutative nor idempotent), v_0 is defined by $a \in v_0(p)$ iff $\vdash_{\mathbf{L}} a \rightarrow p$, the validity of sequents is defined as above and implications are evaluated as follows:

$$
\begin{aligned}
v(B/A) &= \{a \in I \mid (\forall b \in I) \text{ if } b \in v(A), \text{ then } a \oplus b \in v(B)\}, \\
v(A \setminus B) &= \{a \in I \mid (\forall b \in I) \text{ if } b \in v(A), \text{ then } b \oplus a \in v(B)\}.
\end{aligned}
$$

Completeness can then be proved as before. If sequents with empty antecedents are allowed, one may introduce the empty multiset resp. the empty sequence as the empty piece of information 1 and define the validity of sequents $\rightarrow A$ wrt 1.

2.5 Appendix: Possible constraints on 'informational interpretation'

The notion of an informational interpretation of models based on abstract information structures probably cannot be captured by a precise definition. Nevertheless one might want to clarify this notion to a certain extent by imposing some constraints on it. Wrt (minimal or intuitionistic) Kripke models $< I, \sqsubseteq, v_0 >$ e.g. one can say that, given the interpretation of I as a set of information states and $\sqsubseteq$ as the development (or rather the expansion) of these states, the evaluation of L-formulas in Kripke models, the definition of semantic consequence, and the properties of $\sqsubseteq$ can be considered to be *intuitively plausible*. Clearly, rendering the properties of the components of the respective abstract information structures intuitively plausible may be regarded as an essential ingredient of any informational interpretation, and certainly under any truly informational interpretation also the evaluation clauses and the definition of semantic consequence should emerge as plausible. (Otherwise, what would be the explanatory value of the interpretation?) In addition to these basic conditions one might ask for a concrete *intended model*. Moreover, since intuitively *pieces* of information are finitary, 'incomplete' entities, the requirement of *finite representability* may appear to be a natural constraint on the specification of information pieces or states, at least for intended models. In the case of Grzegorczyk's interpretation of IPL e.g. all models taken into account are concrete, the class of researches *is* the class of intended models, and every information piece is a *finite* set. Now, one may ask by virtue of which property a model is an intended model for a given logic $\mathcal{L}$. An obvious answer is "by characterizing $\mathcal{L}$"; in other words every intended model should be canonical (or 'complete'). As far as Urquhart's intrepretation of $R_{\supset}$ is concerned we may note that it produces a model which (i) can arguably be talked about as the intended model and (ii) characterizes the logic and thus the information processing mechanism in question.

Let us summarize the above considerations. If a propositional logic $\mathcal{L}$ in the language L is characterized by a class Γ of models based on certain abstract information structures, then an interpretation of models from Γ will be regarded as an *informational* interpretation of $\mathcal{L}$, if

1. the interpretation renders the properties postulated for the relations, operations, or designated elements plausible (according to certain intuitions);
2. the evaluation clauses for L-formulas and the definition of semantic consequence emerge as plausible (according to intuitions compatible with those referred to under 1);
3. the interpretation provides for a concrete, intended model $\mathcal{M} \in \Gamma$, in which the pieces of information are finitely representable;
4. each intended model $\mathcal{M}$ is a canonical model for $\mathcal{L}$, i.e. $\mathcal{M}$ itself characerizes $\mathcal{L}$.

These suggested criteria are nontrivial. As we have seen in Chapter 1, there exists no canonical research in Grzegorczyk's sense, and hence Grzegorczyk's interpretation violates condition 4.

Chapter 3

Intuitionistic minimal and intuitionistic information processing

The present chapter is devoted to introducing a family of substructural subsystems of MPL together with a related family of subsystems of IPL. The base systems will be 'intuitionistic minimal sequential propositional logic' $MSPL$ resp. 'intuitionistic sequential propositional logic' $ISPL$, i.e. MPL resp. IPL without structural inference rules. The families are unfolded by succesively adding certain such structural rules to sequent-style presentations of $MSPL$ resp. $ISPL$. In both families negation is not an independent operation but defined by means of (the right-searching implication) / or (the left-searching implication) \ and the constant $\bot$. We shall also consider various properties of some of these systems, viz. cut-eliminability, decidability, and interpolation.

3.1 Substructural subsystems of MPL and IPL

Definition 3.1 The rules constituting $MSPL$ are the rules of MPL with the exception of **P**, **C**, and **M**, i.e. $MSPL$ is MPL without structural rules of inference. The rules of $ISPL$ are the rules of $MSPL$ together with ($\bot \rightarrow$), i.e. $ISPL$ is IPL without structural rules of inference.

$\mathcal{D}_{MSPL}(\Pi, X \rightarrow A, S)$, "$\Pi$ is a derivation in $MSPL$ of $X \rightarrow A$ from the finite, possibly empty sequence S of sequent occurrences", resp. $\mathcal{D}_{ISPL}(\Pi, X \rightarrow A, S)$, "$\Pi$ is a derivation in $ISPL$ of $X \rightarrow A$ from the finite, possibly empty sequence S of sequent occurrences", is inductively defined in analogy to $\mathcal{D}_{IPL}(\Pi, X \rightarrow A, S)$ (cf. Appendix 1.5).

By means of (cut) it can easily be verified that instead of the rules ($\rightarrow \top$), ($/ \rightarrow$), ($\backslash \rightarrow$), ($\rightarrow \circ$), ($\wedge \rightarrow$), and ($\rightarrow \vee$) one may equivalently use, respectively:

$$(\top \uparrow) \quad X\top Y \rightarrow A \vdash XY \rightarrow \top;$$

$$(\uparrow /) \quad X \rightarrow (B/A) \vdash XA \rightarrow B;$$

$$(\uparrow \backslash) \quad X \rightarrow (A \backslash B) \vdash AX \rightarrow B;$$

$$(\circ \uparrow) \quad X(A \circ B)Y \rightarrow C \vdash XABY \rightarrow C;$$

$$
\begin{array}{ll}
(\uparrow \wedge) & X \rightarrow (A \wedge B) \vdash X \rightarrow A, \\
 & X \rightarrow (A \wedge B \vdash X \rightarrow B; \\
(\vee \uparrow) & X(A \vee B)Y \rightarrow C \vdash XAY \rightarrow C, \\
 & X(A \vee B)Y \rightarrow C \vdash XBY \rightarrow C.
\end{array}
$$

Adding combinations of the following structural inference rules to the rules of $MSPL$ and $ISPL$ induces families of distinct sublogics of MPL and IPL (cf. [Došen 1988]):

$$
\begin{array}{ll}
\text{permutation } (\mathbf{P}): & XABY \rightarrow C \vdash XBAY \rightarrow C; \\
\text{contraction } (\mathbf{C}): & XAAY \rightarrow B \vdash XAY \rightarrow B; \\
\text{cancellation } (\mathbf{C}'): & XAYAZ \rightarrow B \vdash XAYZ \rightarrow B, \\
 & XAYAZ \rightarrow B \vdash XYAZ \rightarrow B; \\
\text{expansion } (\mathbf{E}): & XAY \rightarrow B \vdash XAAY \rightarrow B; \\
\text{duplication } (\mathbf{E}'): & XAYZ \rightarrow B \vdash XAYAZ \rightarrow B, \\
 & XYAZ \rightarrow B \vdash XAYAZ \rightarrow B; \\
\text{monotonicity } (\mathbf{M}): & XY \rightarrow B \vdash XAY \rightarrow B.
\end{array}
$$

Let Ξ range over $\{MSPL, ISPL\}$, let $\Delta \subseteq \{\mathbf{P}, \mathbf{C}, \mathbf{C}', \mathbf{E}, \mathbf{E}', \mathbf{M}\}$, and let Ξ_Δ denote the extension of Ξ by the structural rules in Δ. Note that $\mathbf{P}$ is derivable in $\Xi_{\{\mathbf{C},\mathbf{M}\}}$:

$$
\dfrac{\dfrac{B \rightarrow B \quad A \rightarrow A}{BA \rightarrow B \circ A} \quad \dfrac{\dfrac{\dfrac{\dfrac{\dfrac{B \rightarrow B}{BA \rightarrow B} \quad \dfrac{A \rightarrow A}{BA \rightarrow A}}{BA \rightarrow B \wedge A}}{B \circ A \rightarrow B \wedge A} \quad \dfrac{\dfrac{\dfrac{A \rightarrow A}{B \wedge A \rightarrow A} \quad \dfrac{B \rightarrow B}{B \wedge A \rightarrow B}}{(B \wedge A)\,(B \wedge A) \rightarrow A \circ B}}{B \wedge A \rightarrow A \circ B}}{B \circ A \rightarrow A \circ B} \quad \dfrac{XABY \rightarrow C}{X(A \circ B)Y \rightarrow C}}{X(B \circ A)Y \rightarrow C}}{XBAY \rightarrow C.}
$$

Clearly, $\mathbf{P}$ is also derivable by means of $\mathbf{M}$ and $\mathbf{C}'$ alone:

$$
\dfrac{\dfrac{XABY \rightarrow B}{XABAY \rightarrow B}}{XBAY \rightarrow B.}
$$

Note that using $(\perp \rightarrow)$, the propositional constant $\mathbf{t}$ can be defined in $ISPL$ as $\perp/\perp$ or as $\perp \backslash \perp$; see e.g.:

$$
\dfrac{\dfrac{\perp \rightarrow (A_n \backslash \ldots \backslash (A_1 \backslash \perp) \ldots) \quad \dfrac{\dfrac{\dfrac{A_1 \rightarrow A_1 \quad \perp \rightarrow \perp}{A_1 (A_1 \backslash \perp) \rightarrow \perp}}{\vdots}}{A_1 \ldots A_n (A_n \backslash \ldots \backslash (A_1 \backslash \perp) \ldots) \rightarrow \perp}}{A_1 \ldots A_n \perp \rightarrow \perp}}{A_1 \ldots A_n \rightarrow \perp/\perp.}
$$

In Figure 3.1, the lattice structure resulting from the addition of the structural rules in Δ to one of the base systems is depicted as a Hasse-diagram, where $\longrightarrow$ denotes proper set-inclusion. The two lattices can be thought of as frameworks offering different options for representing deductive information processing. E.g. in the absence of structural rules the premises are conceived of as *sequences of occurrence*, whereas in the presence of **P** resp. **P**, **C**, and **E** one is dealing with *multisets* resp. *sets* of premises. Different grades of monotonicity of inference are provided by selecting among **E**, **E**′, and **M**. Within the lattices a number of (fragments of) well-known propositional logics can be identifed as deductively equivalent. If empty sequences on the lhs of $\rightarrow$ are excluded, the $\{/, \backslash, \circ\}$-fragment of $MSPL$ is known as the bi-directional, associative Lambek Calculus of Categorial Grammar (see [Lambek 1958]). $ISPL_{\{\mathbf{P}\}}$ turns out to be exactly intuitionistic linear propositional logic without 'exponentials' (cf. [Girard 1987], [Avron 1988], [Troelstra 1992]). The implicational fragments of $MSPL_{\{\mathbf{P,C}\}}$, $MSPL_{\{\mathbf{P,C,E}\}}$, and $MSPL_{\{\mathbf{P,M}\}}$, respectively, can be identified as relevant implicational logic $R_{\supset}$, 'mingle' implicational logic $RM0_{\supset}$, and BCK implicational logic, respectively (cf. e.g. [Dunn 1986], [Ono & Komori 1985]).

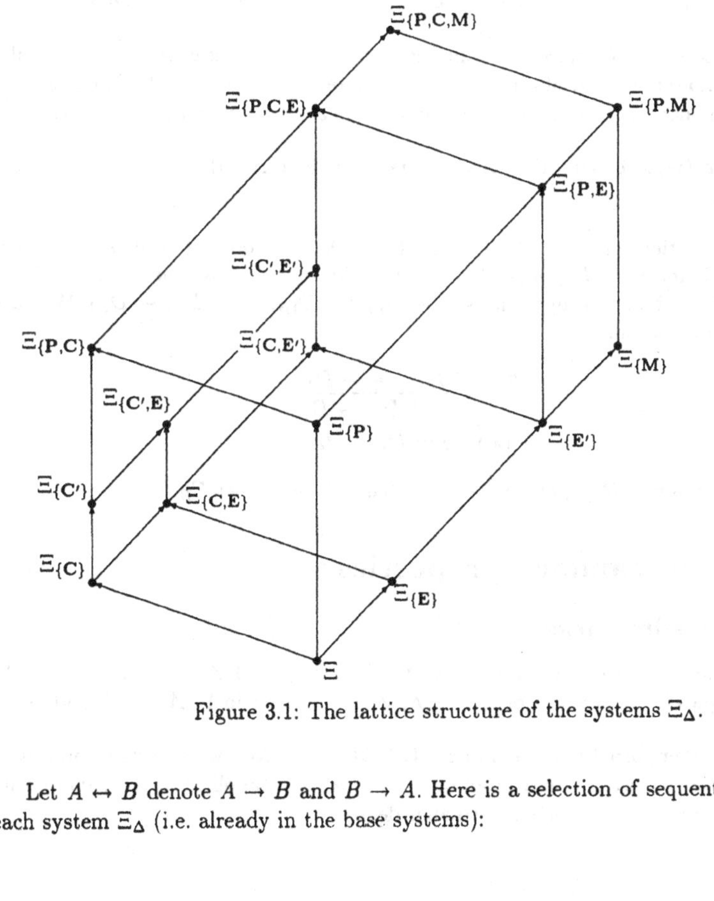

Figure 3.1: The lattice structure of the systems Ξ_{Δ}.

Let $A \leftrightarrow B$ denote $A \rightarrow B$ and $B \rightarrow A$. Here is a selection of sequents provable in each system Ξ_{Δ} (i.e. already in the base systems):

$$A \circ (B \circ C) \leftrightarrow (A \circ B) \circ C,$$
$$A \rightarrow B/(A \backslash B),$$
$$A \rightarrow (B/A) \backslash B,$$
$$A/B \rightarrow (A/C)/(B/C),$$
$$A \backslash B \rightarrow (C \backslash A) \backslash (C \backslash B),$$
$$(A/B)(B/C) \rightarrow (A/C),$$
$$(C \backslash B)(B \backslash A) \rightarrow (C \backslash A),$$
$$(A \backslash B)/C \leftrightarrow A \backslash (B/C),$$
$$(A/B)/C \leftrightarrow A/(C \circ B),$$
$$C \backslash (B \backslash A) \leftrightarrow (B \circ C) \backslash A,$$
$$C \vee (A \wedge B) \rightarrow (C \vee A) \wedge (C \vee B),$$
$$(C \wedge A) \vee (C \wedge B) \rightarrow C \wedge (A \vee B),$$
$$A \leftrightarrow A \circ \top, \quad A \leftrightarrow \top \circ A,$$
$$(\dagger) \quad (A \vee B) \circ C \leftrightarrow (A \circ C) \vee (B \circ C), \quad A \circ (B \vee C) \leftrightarrow (A \circ B) \vee (A \circ C).$$

Let C_A denote an L-formula that contains a certain occurrence of A as a subformula, and let C_B denote the result of replacing this occurrence of A in C by B. The degree of A ($d(A)$) is the number of occurrences of propositional constants and connectives in A.

Theorem 3.2 (replacement) If $\rightarrow A \rightleftharpoons^+ B$ is provable in Ξ_Δ, then so is $\rightarrow C_A \rightleftharpoons^+ C_B$.

PROOF By induction on $l = d(C_A) - d(A)$. If $l = 0$, the proof is trivial. Assume that the claim holds for every $l \leq m$, and $l = m + 1$. We consider just one case, viz. $C_A = D_{1A} \backslash D_2$. By the induction hypothesis, $\vdash \rightarrow D_{1A} \rightleftharpoons^+ D_{1B}$, i.e. $\vdash D_{1A} \leftrightarrow D_{1B}$. We have the following derivation:

$$\cfrac{\cfrac{D_{1B} \rightarrow D_{1A} \quad D_2 \rightarrow D_2}{D_{1B} \quad (D_{1A} \backslash D_2) \rightarrow D_2}}{D_{1A} \backslash D_2 \rightarrow D_{1B} \backslash D_2}$$

Similarly we obtain $\vdash D_{1B} \backslash D_2 \rightarrow D_{1A} \backslash D_2$. Thus, $\vdash \rightarrow C_A \rightleftharpoons^+ C_B$. □

3.2 Some standard properties

3.2.1 Cut-elimination

Let the degree of an application $Y \rightarrow A \; XAZ \rightarrow B \vdash XYZ \rightarrow B$ of (cut) be the number of separate occurrences of $/$, $\backslash$, $\circ$, $\wedge$, $\vee$, $\bot$, $\top$, and $\mathbf{t}$ in Y, A, X, Z, and B.

Lemma 3.3 Every proof of $X \rightarrow A$ in Ξ_Θ, $\Theta \subseteq \{\mathbf{P}, \mathbf{C}, \mathbf{C}', \mathbf{M}\}$, with a single application of (cut) can be converted into a proof of $X \rightarrow A$ with no application of (cut) or with one or two applications of (cut) of a smaller degree.

PROOF By a standard distinction between possible proofs of the premise sequents of (cut) (cf. e.g. [Lambek 1958] and [Došen 1988]). Case 1: One of the premise sequents of (cut) is of the form $A \to A$. Then the conclusion of (cut) is identical with the remaining premise (sequent), which is proved without applying (cut). Case 2: If the left premise of (cut) is $\to \top$, replace in the proof of the right premise $X\top Z \to A$ every occurrence of $\top$ on the lhs of sequent arrows. The result is a (cut)-free proof of $XZ \to A$. Case 3: The last step in the proof of the left or the right premise of (cut) is the application of an operational rule, such that the connective or constant introduced is not the main connective of the cut-formula (i.e. the formula which is eliminated by applying (cut)) resp. is not the cut-formula. We must deal with all possible cases and show that an application of the operational rule in question need never immediately precede an application of (cut). Consider by way of example the following conversions:

$$\left[\dfrac{\dfrac{\dfrac{\Pi_1}{Y_1 \to C} \quad \dfrac{\Pi_2}{X_1 D Z_1 \to A}}{X_1(D/C)Y_1 Z_1 \to A} \quad \dfrac{\Pi_3}{XAZ \to B}}{XX_1(D/C)Y_1 Z_1 Z \to B}\right] \text{ is converted into } \left[\dfrac{\dfrac{\Pi_1}{Y_1 \to C} \quad \dfrac{\dfrac{\Pi_2}{X_1 D Z_1 \to A} \quad \dfrac{\Pi_3}{XAZ \to B}}{XX_1 D Z_1 Z \to B}}{XX_1(D/C)Y_1 Z_1 Z \to B}\right];$$

$$\left[\dfrac{\dfrac{\Pi_1}{Y \to A} \quad \dfrac{\dfrac{\Pi_2}{XAZC \to B}}{XAZ \to (B/C)}}{XYZ \to (B/C)}\right] \text{ is converted into } \left[\dfrac{\dfrac{\dfrac{\Pi_1}{Y \to A} \quad \dfrac{\Pi_2}{XAZC \to B}}{XYZC \to B}}{XYZ \to (B/C)}\right].$$

Case 4: The last step in the proof of both premises of (cut) is the application of an operational rule apart from $(\bot \to)$ and $(\to \mathbf{t})$, such that the connective or constant introduced is the main connective of the cut-formula resp. is the cut-formula. We have to consider all possible cases and show that an application of the operational rule in question is superfluous: we can instead use one or two applications of (cut) of a smaller degree. For instance,

$$\left[\dfrac{\dfrac{\dfrac{\Pi_1}{XA \to B}}{X \to (B/A)} \quad \dfrac{\dfrac{\Pi_2}{Y_1 \to A} \quad \dfrac{\Pi_3}{X_1 B Z \to C}}{X_1(B/A)Y_1 Z \to C}}{X_1 X Y_1 Z \to C}\right] \text{ is converted into } \left[\dfrac{\dfrac{\dfrac{\Pi_2}{Y_1 \to A} \quad \dfrac{\Pi_1}{XA \to B}}{XY_1 \to B} \quad \dfrac{\Pi_3}{X_1 B Z \to C}}{X_1 X Y_1 Z \to C}\right].$$

Case 5: Both premises of (cut) are instantiations of $(\bot \to)$ resp. $(\to \mathbf{t})$:

$$\dfrac{Y \to \mathbf{t} \quad X\mathbf{t}Z \to \mathbf{t}}{XYZ \to \mathbf{t}} \quad \text{is converted into} \quad XYZ \to \mathbf{t};$$

$$\dfrac{Y_1 \bot Y_2 \to \bot \quad X \bot Z \to A}{XY_1 \bot Y_2 Z \to A} \quad \text{is converted into} \quad XY_1 \bot Y_2 Z \to A.$$

Case 6: The last step in the proof of the left or the right premise of (cut) is the application of a structural rule $\mathsf{R} \in \Theta$. We show that an application of R need never immediately precede an application of (cut). (a) The last step in the proof of the left premise of (cut) is an application of R, $\mathsf{R} = \mathbf{P}$ or $\mathbf{M}$:

$$\left[\dfrac{\dfrac{\dfrac{\Pi_1}{Y_1 \to A}}{Y_2 \to A}\,\mathsf{R} \quad \dfrac{\Pi_2}{Z_1 A Z_2 \to B}}{Z_1 Y_2 Z_2 \to B}\right] \text{ is converted into } \left[\dfrac{\dfrac{\dfrac{\Pi_1}{Y_1 \to A} \quad \dfrac{\Pi_2}{Z_1 A Z_2 \to B}}{Z_1 Y_1 Z_2 \to B}}{Z_1 Y_2 Z_2 \to B}\,\mathsf{R}\right].$$

$\mathsf{R} = \mathbf{C}$ (the case $\mathsf{R} = \mathbf{C}'$ is analogous):

$$\left[\frac{\dfrac{\dfrac{\Pi_1}{Y_1 \to A}}{Y_2 \to A}\,\mathsf{R} \qquad \dfrac{\Pi_2}{Z_1 A Z_2 \to B}}{Z_1 Y_2 Z_2 \to B}\right] \text{ is converted into } \left[\dfrac{\dfrac{\dfrac{\Pi_1}{Y_1 \to A} \quad \dfrac{\Pi_2}{Z_1 A Z_2 \to B}}{Z_1 Y_1 Z_2 \to B}}{Z_1 Y_2 Z_2 \to B}\,\mathsf{R}\right].$$

(b) The last step in the proof of the right premise of (*cut*) an application of R. (i) The cut-formula has neither been introduced by the application of **M** nor has it been contracted resp. canceled by the application of **C** resp. **C**′. For $\mathsf{R} = \mathbf{C}$ e.g. one obtains the following conversion step:

$$\left[\frac{\dfrac{\Pi_1}{Y \to A} \qquad \dfrac{\dfrac{\Pi_2}{Z_1 A Z_2 \to B}}{Z_3 A Z_2 \to B}\,\mathsf{R}}{Z_3 Y_1 Z_2 \to B}\right] \text{ is converted into } \left[\dfrac{\dfrac{\dfrac{\Pi_1}{Y \to A} \quad \dfrac{\Pi_2}{Z_1 A Z_2 \to B}}{Z_1 Y_1 Z_2 \to B}}{Z_3 Y_1 Z_2 \to B}\,\mathsf{R}\right].$$

(ii) The last step in the proof of the right premise of (*cut*) is an applcation of $\mathsf{R} \in \{\mathbf{C}, \mathbf{C}', \mathbf{M}\}$ such that the cut-formula has been introduced by the application of **M** resp. it has been contracted (canceled) by the application of **C** (**C**′). We present two conversion steps, one for **C**′ and one for **M**:

$$\left[\frac{\dfrac{\Pi_1}{Y_1 \to A} \qquad \dfrac{\dfrac{\Pi_2}{Z_1 A Z_2 A Z_3 \to B}}{Z_1 Z_2 A Z_3 \to B}}{Z_1 Z_2 Y_1 Z_3 \to B}\right] \text{ is converted into } \left[\begin{array}{c}\dfrac{\dfrac{\Pi_1}{Y_1 \to A} \qquad \dfrac{\dfrac{\Pi_1}{Y_1 \to A} \quad \dfrac{\Pi_2}{Z_1 A Z_2 A Z_3 \to B}}{Z_1 Y_1 Z_2 A Z_3 \to B}}{Z_1 Y_1 Z_2 Y_1 Z_3 \to B} \\ \vdots \\ \hline Z_1 Z_2 Y_1 Z_3 \to B\end{array}\right];$$

$$\left[\frac{\dfrac{\Pi_1}{Y_1 \to A} \qquad \dfrac{\dfrac{\Pi_2}{Z_1 Z_2 \to B}}{Z_1 A Z_2 \to B}}{Z_1 Y Z_2 \to B}\right] \text{ is converted into } \left[\begin{array}{c}\dfrac{\Pi_2}{Z_1 Z_2 \to B} \\ \vdots \\ \hline Z_1 Y_1 Z_2 \to B\end{array}\right]. \ \square$$

Theorem 3.4 (cut-elimination) Applications of (*cut*) can be eliminated from proofs in Ξ_Θ, where $\Theta \subseteq \{\mathbf{P}, \mathbf{C}, \mathbf{C}', \mathbf{M}\}$.

PROOF Since no application of (*cut*) has a negative degree, applications of (*cut*) can be eliminated from proofs in Ξ_Θ by applying the conversion algorithm described in the proof of the previous lemma, starting from the top of proofs, proceeding from left to right. □[1]

Applications of (*cut*) cannot be eliminated from proofs in Ξ_Δ if **E** or **E**′ are in Δ, but **M** is not. This holds already for the implicational fragments. The following counter-example to cut-eliminability for the implicational fragment of $\Xi_{\{\mathbf{E}\}}$ is due to Kosta Došen (personal communcation): $p_1(p_2/(p_1 \setminus p_2)) \to (p_2/(p_1 \setminus p_2))$. Došen also pointed out that this sequent is not a counterexample to cut-eliminability for proofs in

[1] A strengthening of cut-elimination which will not be considered here is strong cut-elimination, saying that every sequence of conversion steps terminates in a (*cut*)-free proof (cf. e.g. [Roorda 1991]).

the implicational fragment of $\Xi_{\{\mathbf{P},\mathbf{E}\}}$. Such a counter-example has been found by Dirk Roorda (personal communication): $p_1(p_1 \backslash p_2)(p_2 \backslash p_3)(p_1 \backslash p_3) \to p_3$. A counterexample for the implicational fragment of $\Xi_{\{\mathbf{P},\mathbf{C},\mathbf{E}\}}$ is presented in the following subsection on decidabilty. The problem with **E** and **E'** is that certain applications of (*cut*) cannot be pushed upwards:

$$\frac{XAZ \to B \quad \dfrac{Y \to A}{XAAZ \to B}}{XAYZ \to B,}$$

where A does not occur in X or Y. Note that in the presence of (*cut*) instead of the rule **E** one may equivalently use the rule

$$\text{mingle } (\mathbf{MI}): \; X \to A \quad Y \to A \vdash XY \to A.$$

Thus, $\Xi_{\{\mathbf{E}\}} = \Xi_{\{\mathbf{MI}\}}$. In the absence of **M**, also **MI** blocks cut-eliminability. Here we have problematic cases like the following:

$$\left[\frac{\dfrac{\dfrac{\Pi_1}{X\to A}\quad\dfrac{\Pi_2}{Y\to A}}{XY\to A}\quad \dfrac{\Pi_3}{X_1AZ\to B}}{X_1XYZ\to B}\right] \text{ is converted into}$$

$$\left[\begin{array}{c} \dfrac{\dfrac{\Pi_1}{X\to A}\quad\dfrac{\Pi_3}{X_1AZ\to B}}{X_1XZ\to B} \qquad \dfrac{\dfrac{\Pi_2}{Y\to A}\quad\dfrac{\Pi_3}{X_1AZ\to B}}{X_1YZ\to B}\\ \vdots \qquad\qquad\qquad \vdots \\ \dfrac{X\to(\ldots(A_n\backslash\ldots(A_1\backslash(\ldots(B/B_m)/\ldots/B_1))\ldots) \quad Y\to(\ldots(A_n\backslash\ldots(A_1\backslash(\ldots(B/B_m)/\ldots/B_1))\ldots)}{XY\to(\ldots(A_n\backslash\ldots(A_1\backslash(\ldots(B/B_m)/\ldots/B_1))\ldots)} \\ \vdots \\ \overline{X_1XYZ\to B,} \end{array}\right]$$

where $X_1 = A_1 \ldots A_n$ and $Z = B_1 \ldots B_m$. The steps from $XY \to (\ldots(A_n \backslash \ldots (A_1 \backslash (\ldots(B/B_m)/\ldots/B_1))\ldots)$ to $X_1XYZ \to B$ involve applications of (*cut*). In the converted proof we still have an application of **MI** that is immediately followed by an application of (*cut*), and unfortunately this constellation may *loop*, as can be tested with the following example:

$$\frac{\dfrac{\dfrac{\Pi_1}{(p_1\backslash p_2)(p_2\backslash p_3)\to p_1\backslash p_3}\quad p_1\backslash p_3\to p_1\backslash p_3}{(p_1\backslash p_2)(p_2\backslash p_3)(p_1\backslash p_3)\to p_1\backslash p_3}\quad \dfrac{\Pi_2}{p_1(p_1\backslash p_3)\to p_3}}{p_1(p_1\backslash p_2)(p_2\backslash p_3)(p_1\backslash p_3)\to p_3.}$$

The above mentioned counterexamples to cut-eliminability remain counterexamples, if **E** is replaced by **MI**. (See the decision procedures in the following subsection).[2]

Cut-elimination has a number of useful immediate consequences.

[2] Nuel Belnap pointed out to me that in the presence of (*cut*), **MI** is interreplaceable with the rule $X \to A \quad Y \to A \quad Z_1AZ_2 \to B \vdash Z_1XYZ_2 \to B$. This rule allows to eliminate (*cut*) also in the absence of **M**.

Definition 3.5 Let $\mathcal{L}_1$, $\mathcal{L}_2$ be logics (with sequent calculus presentations) such that the language L_2 of $\mathcal{L}_2$ extends the language L_1 of $\mathcal{L}_1$. We say that $\mathcal{L}_1$ is a conservative subsystem of $\mathcal{L}_2$ and $\mathcal{L}_2$ is a conservative extension of $\mathcal{L}_1$ iff for every L_1-formula A and every finite sequence X of L_1-formula occurrences:

$\vdash_{\mathcal{L}_1} X \to A$ iff $\vdash_{\mathcal{L}_2} X \to A$.

Corollary 3.6 (i) (subformula property) If $\vdash_{\Xi_\Theta} A_1 \dots A_n \to A$, then there is a proof of this sequent in Ξ_Θ in which only subformulas of $A_1, \dots A_n$, and A occur;

(ii) Each subsystem obtained from Ξ_Θ by dropping all rules for a certain constant or connective is a conservative subsystem of Ξ_Θ;

(iii) If $\vdash_{\Xi_\Theta} \to A$, then there is a proof of this sequent in Ξ_Θ in which the last step is the application of an operational rule introducing a connective on the rhs of $\to$;

(iv) (consistency) $\to \bot$ cannot be proved in Ξ_Θ;

(v) (disjunction property) If $A \vee B$ is provable in Ξ_Θ (i.e. $\vdash_{\Xi_\Theta} \to A \vee B$), then A is provable or B is provable.

Corollary 3.7 *Tertium non datur* in both forms $A \vee \neg^r A$ and $A \vee \neg^l A$ is not a theorem of Ξ_Θ.

PROOF By the disjunction property, e.g. $\not\vdash p_1 \vee \neg^r p_1$, since neither $\vdash p_1$ nor $\vdash \neg^r p_1$. □

Theorem 3.8 Ξ_{Δ_1} is not an n-valued logic, $1 \leq n < \omega$.

PROOF The proof is essentially the same as Gödel's proof that IPL is not a finitely many-valued logic (see [Gödel 1932]). Since in Ξ_Θ, $\vdash A \leftrightarrow B$ iff $\vdash \to A \rightleftharpoons^+ B$, and $\vdash \to A \vee B$ iff $\vdash \to A$ or $\vdash \to B$, every adequate n-valued truth-table with designated value 1 for $\rightleftharpoons^+$ resp. $\vee$ must look like this

$\rightleftharpoons^+$	1	...	n
1	1		
$\vdots$		$\ddots$	
n			1

$\vee$	1	...	n
1	1	...	1
$\vdots$	$\vdots$		
n	1		

Now, suppose that Ξ_Θ is an n-valued logic with designated value 1. Then any disjunction of all equivalences $p_i \rightleftharpoons^+ p_j$ where $i < j \leq n+1$ is a theorem of Ξ_Δ, since at least one equivalence $p_i \rightleftharpoons^+ p_j$ is evaluated as 1. By the disjunction property, one of the disjuncts is a theorem of Ξ_Θ, quod non. □

3.2.2 Decidability

In the base systems $MSPL$ and $ISPL$ each sequent rule apart from (*cut*) introduces complexity (viz. one additional occurrence of a connective or propositional constant) in passing from the premise sequent(s) to the conclusion. Also the structural inference rules **P**, **E**, **E′**, **MI**, and **M** do not reduce complexity. This observation leads to an obvious decision procedure for certain systems which admit of (*cut*)-elimination.

Theorem 3.9 Provability of sequents in Ξ_{Δ_1} is decidable, $\Delta_1 \subseteq \{\mathbf{P}, \mathbf{M}\}$.

PROOF Every sequent provable in Ξ_{Δ_1} is provable without applying (*cut*). Now, take any sequent s and draw a line from s to each premise sequent s_0 or sequence of premise sequents $s_1 s_2$ such that s can be derived from s_0 or $s_1 s_2$. If **P** resp. **M** is present, draw also a line from s to every result of permuting the antecedent of s resp. to every sequent s' from which s is derivable by applying **M**. Iterate this process so as to obtain the *complete proof-search tree* of s wrt Ξ_{Δ_1}. Since the number of sequent rules is finite, s is finite, and no rule reduces complexity, the complete proof-search tree of s is finite and thus a decision procedure exists. □

Corollary 3.10 Ξ_{Δ_2} minus (*cut*) is decidable, $\Delta_2 \subseteq \{\mathbf{P}, \mathbf{E}, \mathbf{E}', \mathbf{MI}, \mathbf{M}\}$.

PROOF Obvious. In the presence of **MI** we neglect in the proof-search tree for a sequent $X \to A$ the branches leading from $X \to A$ to $X \to A \quad \to A$ or to $\to A \quad X \to A$. □

It can now easily be shown that e.g. neither $(C \vee A) \wedge (C \vee B) \to C \vee (A \wedge B)$ nor $C \wedge (A \vee B) \to (C \wedge A) \vee (C \wedge B)$ is provable in $ISPL$. Moreover we can verify that the sequents $p_1(p_2/(p_1 \setminus p_2)) \to (p_2/(p_1 \setminus p_2))$ resp. $p_1(p_1 \setminus p_2)(p_2 \setminus p_3)(p_1 \setminus p_3) \to p_3$ in fact are counterexamples to cut-eliminability in $\Xi_{\{\mathbf{E}\}} = \Xi_{\{\mathbf{MI}\}}$ resp. $\Xi_{\{\mathbf{P},\mathbf{E}\}} = \Xi_{\{\mathbf{P},\mathbf{MI}\}}$.

Complexity-reducing rules like **C** and **C**$'$ may pose a problem for proving decidability, since in their presence the method of constructing complete proof-search trees does not guarantee the finiteness of the search-space. In what follows we analyse and apply to a wider class of substructural logics a technique used by Kripke to prove decidability of $\vdash_{R_{\supset}}$, i.e. the implicational fragment of $MSPL_{\{\mathbf{P},\mathbf{C}\}}$. This will allow us to include **C** and **C**$'$. We shall draw on the clear presentation of Kripke's argument in [Dunn 1986]; a proof of Kripke's Lemma (which is a form of Kruskal's Theorem in graph theory, see [van Benthem 1991]) can be found in [Anderson & Belnap 1975].

Theorem 3.11 Provability of sequents in Ξ_Θ is decidable, $\Theta \subseteq \{\mathbf{P}, \mathbf{C}, \mathbf{C}', \mathbf{M}\}$.

PROOF We will give a proof for the case that $\mathbf{C} \in \Theta$ which can be converted into a proof for the case that $\mathbf{C}' \in \Theta$ by replacing 'contraction' resp. '**C**' resp. 'contracting' throughout by 'cancellation' resp. '**C**$'$' resp. 'cancelling'. We say that $X' \to A$ is a *contraction* of $X \to A$ iff $X' \to A$ can be derived from $X \to A$ by repeated applications of **C**. In a first step we shall replace Ξ_Θ by an equivalent calculus Ξ'_Θ which suits for the proof of decidability. The idea is to get rid of **C** by building into the operational rules a restricted amount of contraction. This is achieved by allowing a contraction of the conclusion of an operational rule only in so far as the contraction cannot be obtained by applying the operational rule in question after first contracting the premises. The following modified rules will suffice:

$$(/\to)^0 \quad Y \to A \quad XBZ \to C \vdash [X(B/A)YZ] \to C,$$
$$(\setminus\to)^0 \quad Y \to A \quad XBZ \to C \vdash [XY(A \setminus B)YZ] \to C,$$
$$(\top\to)^0 \quad XY \to A \vdash [X\top Y] \to A,$$
$$(\to\circ)^0 \quad X \to A \quad Y \to B \vdash [XY] \to A \circ B,$$
$$(\circ\to)^0 \quad XABY \to C \vdash [X(A \circ B)Y] \to C.$$

$$(\wedge \rightarrow)^0 \quad XAY \rightarrow C \vdash [X(A \wedge B)Y] \rightarrow C,$$
$$XBY \rightarrow C \vdash [X(A \wedge B)Y] \rightarrow C,$$
$$(\vee \rightarrow)^0 \quad XAY \rightarrow C \quad XBY \rightarrow C \vdash [X(A \vee B)Y] \rightarrow C,$$

where $[X(B/A)YZ]$ resp. $[XY(A \setminus B)Z]$ is the contraction of $X(B/A)YZ$ resp. $XY(A \setminus B)Z$ such that

(i) (B/A) resp. $(A \setminus B)$ occurs only 0, 1, or 2 times fewer than in $X(B/A)YZ$ resp. $XY(A \setminus B)Z$,

(ii) any formula other than (B/A) resp. $(A \setminus B)$ occurs only 0 or 1 time fewer,

where $[X\top Y] \rightarrow A$ is the contraction of $X\top Y \rightarrow A$ such that $\top$ occurs only 0 or 1 time fewer than in $X\top Y \rightarrow A$,

where $[XY] \rightarrow A \circ B$ is the contraction of $XY \rightarrow A \circ B$ such that any formula in XY occurs only 0 or 1 time fewer in $[XY]$ than in XY,

and where $[X(A\nabla B)Y] \rightarrow C$ is the contraction of $X(A\nabla B)Y \rightarrow C$ such that ∇ occurs only 0 or 1 time fewer than in $X(A\nabla B)Y \rightarrow C$, $\nabla \in \{\circ, \wedge, \vee\}$.

REMARK In restriction (i) we have to take into account using C twice because of proofs like

$$\dfrac{\dfrac{\dfrac{\vdots \qquad \rightarrow \top \quad (\top \setminus B)B(\top \setminus B) \rightarrow (\top \setminus B) \circ (B \circ (\top \setminus B))}{(\top \setminus B)(\top \setminus B)(\top \setminus B) \rightarrow (\top \setminus B) \circ (B \circ (\top \setminus B))}}{(\top \setminus B)(\top \setminus B) \rightarrow (\top \setminus B) \circ (B \circ (\top \setminus B))}}{(\top \setminus B) \rightarrow (\top \setminus B) \circ (B \circ (\top \setminus B)).}$$

Restriction (ii) is imposed in view of proofs like

$$\dfrac{\dfrac{A \rightarrow A \quad \dfrac{A \rightarrow A \quad B \rightarrow B}{AB \rightarrow A \circ B}}{AA(A \setminus B) \rightarrow A \circ B}}{A(A \setminus B) \rightarrow A \circ B.}$$

Lemma 3.12 (Curry) If $X \rightarrow A$ is provable in Ξ'_Θ in n steps and s is a contraction of $X \rightarrow A$, then s is provable in Ξ'_Θ in m steps, $m \leq n$.

PROOF By induction on the number of steps in proofs in Ξ'_Θ. Consider just one example. Suppose that $X \rightarrow A$ has been proved in m_1 steps and $Y \rightarrow B$ has been proved in m_2 steps. Then $[XY] \rightarrow A \circ B$ has a proof in $n = m_1 + m_2 + 1$ steps. Now any contraction $[XY]' \rightarrow A \circ B$ of $[XY] \rightarrow A \circ B$ can be obtained by first contracting $X \rightarrow A$, $Y \rightarrow B$ and then applying $(\rightarrow \circ)^0$. So, just use the induction hypothesis. □

Definition 3.13 A proof in a sequent calculus is said to be irredundant if it contains no branch with a sequent s' below a sequent s such that s' is a contraction of s.

Corollary 3.14 (i) Provability of sequents in Ξ_Θ coincides with provability of sequents in Ξ'_Θ. (ii) Every sequent provable in Ξ'_Θ has an irredundant proof in Ξ'_Θ.

By the previous corollary, complete proof-search trees wrt Ξ'_Θ can be constructed to be irredundant. Note that whereas in the complete proof-search tree for a sequent s wrt Ξ_Θ each node has infinitely many immediate successors, in the case of Ξ'_Θ the number of immediate successors is *finite*. Using

Lemma 3.15 (König) A tree is finite iff each node has only a finite number of immediate successors and each branch is finite,

in order to prove decidability it remains to be shown that irredundant complete proof-search trees wrt Ξ'_Θ have the 'finite branch property'. Note that Ξ'_Θ has the subformula property. As we will see, the finite branch property follows from what Dunn calls Kripke's Lemma and the subformula property. Thus, although some of the operational rules of Ξ'_Θ are complexity-decreasing, irredundant complete proof-search trees wrt Ξ'_Θ are finite and therefore provability of sequents in Ξ'_Θ is decidable.

Definition 3.16 Two sequents $X \to A$, $X' \to A$ are called cognate iff exactly the same formulas occur in X and X'. The class of sequents cognate to a given sequent s is called the cognation class of s. A sequence $s_0, s_1, \ldots$ of cognate sequents is said to be irredundant iff for no s_i, s_j with $i < j$, s_i is a contraction of s_j.

Lemma 3.17 (Kripke) Every irredundant sequence of cognate sequents is finite.

By the subformula property, the number of cognation classes in any proof in Ξ'_Θ is finite. By Kripke's Lemma, only finitely many members of each cognation class occur in a branch of an irredundant complete proof-search tree wrt Ξ'_Θ. Therefore each such branch is finite and hence each irredundant complete proof-search tree wrt Ξ'_Θ is finite. □

Corollary 3.18 The (cut)-free parts of $\Xi_{\{\mathbf{P},\mathbf{C},\mathbf{E}\}}$, $\Xi_{\{\mathbf{P},\mathbf{C},\mathbf{MI}\}}$, and $\Xi_{\{\mathbf{C},\mathbf{E}'\}}$ are decidable.

Using the (extremely non-constructive) decision procedure of constructing irredundant complete proof-search trees it can be seen that the sequent $s_1 = (p_1 \backslash p_2)(p_2 \backslash p_3)(p_1 \backslash p_3) \to ((p_1 \backslash p_3) \backslash p_4) \backslash p_4$ has no (cut)-free proof in $\Xi_{/,\backslash\{\mathbf{P},\mathbf{C},\mathbf{E}\}} = \Xi_{/,\backslash\{\mathbf{P},\mathbf{C},\mathbf{MI}\}}$[3] and that $s_2 = (p_1 \backslash p_2)(p_2 \backslash p_3)(p_1 \backslash p_3) \to p_4/((p_1 \backslash p_3) \backslash p_4)$ has no (cut)-free proof in $\Xi_{\{\mathbf{C},\mathbf{E}'\}}$. The only reasonable approach to a proof-search for s_2 in $\Xi'_{\{\mathbf{C},\mathbf{E}'\}}$, which is the equivalent version of $\Xi_{\{\mathbf{C},\mathbf{E}'\}}$ obtained by the method of the previous proof, is:

$$\dfrac{\dfrac{(p_1 \backslash p_3) \to (p_1 \backslash p_3) \quad \dfrac{?}{(p_1 \backslash p_2)(p_2 \backslash p_3)p_4 \to p_4}}{(p_1 \backslash p_2)(p_2 \backslash p_3)(p_1 \backslash p_3)((p_1 \backslash p_3) \backslash p_4) \to p_4}}{(p_1 \backslash p_2)(p_2 \backslash p_3)(p_1 \backslash p_3) \to p_4/((p_1 \backslash p_3) \backslash p_4).}$$

In a proof-search for s_1 in $\Xi'_{\{\mathbf{P},\mathbf{C},\mathbf{E}\}}$ a reasonable trial besides an analogue of the above attempted proof is (surpressing permutations):

[3] Note that this contradicts a claim in [Tamura 1971]. Moreover, after Tamura's paper appeared, R. Meyer has shown that mingle implicational logic $RM0_\supset$, contrary to what Tamura assumed, fails to be the implicational fragment of the relevance logic **RM** (see e.g. [Dunn 1986, p. 131]).

$$\dfrac{\dfrac{(p_1 \backslash p_2)(p_2 \backslash p_3) \to (p_1 \backslash p_3) \quad \dfrac{?}{(p_1 \backslash p_2) p_4 \to p_4}}{(p_1 \backslash p_2)(p_2 \backslash p_3)(p_1 \backslash p_3)((p_1 \backslash p_3) \backslash p_4) \to p_4}}{(p_1 \backslash p_2)(p_2 \backslash p_3)(p_1 \backslash p_3) \to ((p_1 \backslash p_3) \backslash p_4) \backslash p_4.}$$

In $\Xi'_{\{\mathbf{P},\mathbf{C},\mathbf{MI}\}}$ one trial to prove s_1 would be:

$$\dfrac{\dfrac{\dfrac{?}{(p_1 \backslash p_2)(p_2 \backslash p_3) \to p_4} \quad (p_1 \backslash p_2) p_4 \to p_4}{(p_1 \backslash p_3)((p_1 \backslash p_3) \backslash p_4) \to p_4}}{(p_1 \backslash p_2)(p_2 \backslash p_3)(p_1 \backslash p_3) \to ((p_1 \backslash p_3) \backslash p_4) \backslash p_4.}$$

3.2.3 Interpolation

The interpolation property can be thought of as a relevance criterion for derivability: the Craig-interpolation theorem for classical propositional logic says that if a $A \supset B$ is provable then there exists a formula C built up only by propositional variables occuring in *both* A and B such that $A \supset C$ and $C \supset B$ are provable. Craig-interpolation can be strengthened to Lyndon-interpolation by requiring the interpolant C to be sensitive to polarities, i.e. 'positive' and 'negative' occurrences, in the following sense: each propositional variable ocurring positively resp. negatively in C occurs positively resp. negatively in both A and B. The Lyndon-interpolation theorem holds for IPL, too; to prove it by proof-theoretic means one needs cut-eliminability and a strenthening of the induction hypothesis thought out by Schütte [1962].

The notions of positive and negative occurrence of propositional variables in L-formulas, sequents built up from L-formulas and sequences of L-formula occurrences, are defined as follows:

Definition 3.19 A propositional variable p occurs positively in the scope of an even number of occurrences of negation signs (not distinguishing between $\neg^r$ and $\neg^r$); it occurs negatively in the scope of an uneven number of occurrences of negation signs. A positive resp. negative occurrence of p in A remains positive resp. negative in $A \wedge B$, $B \wedge A$, $A \circ B$, $B \circ A$, $A \vee B$, $B \vee A$ and $X \to A$; the polarity of p in A is reversed in $\neg^r A$, $\neg^l A$, $A \backslash B$, B/A, and $XAY \to B$. Let $\mathring{X} = A_1 \circ \ldots \circ A_n$, if $X = A_1 \ldots A_n\,(n > 1)$, and let $\mathring{X} = A$, if $X = A$. A propositional variable occurs positively resp. negatively in X iff it occurs positively resp. negatively in $\mathring{X}$. (Of course, no propositional variable occurs in the empty sequence.) Let $pos(X)$ resp. $neg(X)$ denote the set of propositional variables that occur positively resp. negatively in X. A reversal of polarities in a sequence X is indicated by $\bar{X}$.

Theorem 3.20 (interpolation) If $\vdash_{\Xi_\Theta} X \to A$, then there is an L-formula C such that $\vdash_{\Xi_\Theta} X \to C$, $\vdash_{\Xi_\Theta} C \to A$, $pos(C) \subseteq (pos(X) \cap pos(A))$, and $neg(C) \subseteq (neg(X) \cap neg(A))$, $\Theta \subseteq \{\mathbf{P}, \mathbf{C}, \mathbf{C}', \mathbf{M}\}$.

PROOF By induction on $\vdash_{\Xi_\Theta}$. For the proof we shall strengthen the induction hypothesis except for the case of **P** and **C**′. The method we use is due to Schütte [1962]; we shall adopt the presentation in [Roorda 1991] used to prove interpolation for certain fragments of linear logic. Note that the only cases in which Schütte's method is really needed are $(\to /)$ and $(\to \backslash)$.

The idea behind Schütte's method is this: Think of the sequent arrow as a partition marker in a finite sequence of formula occurrences and think of an interpolant as a connecting link between two partitioned sequences. Instead of just the particular partition which is marked by the sequent arrow one might want to consider every possible partition and look for connecting links between the separated parts. We shall effect such partitions of a sequent $X \to A$ by selecting a subsequence of XA. Selections will be indicated by means of underlining. We say that a sequent $X \to A$ satisfies the induction hypothesis if

IND for all $Y_1\underline{Z}Y_2$ such that $X = Y_1\underline{Z}Y_2$ there is a C such that $\vdash Z \to C$, $\vdash Y_1CY_2 \to A$, $pos(C) \subseteq (pos(Z) \cap pos(\overline{Y_1}\,\overline{Y_2}\ A))$, and $neg(C) \subseteq (neg(Z) \cap neg(\overline{Y_1}\,\overline{Y_2}\ A))$.

We may now consider the rules of Ξ_Θ. In each case to be distinguished, apart from those for **P** and **C**′, first a part of the conclusion is selected. This induces a suitable selection in the premise sequents with interpolants provided by IND. From these interpolants we obtain an interpolant for the conclusion. That it is in fact an interpolant can readily be verified. Interpolants will be specified at the sequent arrow.

(id): $\vdash \underline{A} \xrightarrow{A} A$, $\vdash \underline{<>}A \xrightarrow{A/A} A$, $\vdash A\underline{<>} \xrightarrow{A\backslash A} A$, $\vdash \underline{<>}A \xrightarrow{\top} A$, $\vdash A\underline{<>} \xrightarrow{\top} A$.

$(\bot \to)$: $\vdash \underline{X}\bot Y \xrightarrow{\mathbf{t}} A$, $\vdash X\bot\underline{Y} \xrightarrow{\mathbf{t}} A$. For the remaining selections choose $\bot$ as interpolant.

$(\to \mathbf{t})$: For every selection choose **t** as interpolant.

$(\to \top)$: $\vdash \underline{<>} \xrightarrow{\top} \top$

$(\top \to)$: for each selection choose the interpolant from the premise sequent. If e.g. $\top Y$ is selected in the conclusion, select Y in the premise.

$(\to /)$: $\underline{X}A \xrightarrow{C} B \vdash \underline{X} \xrightarrow{C} B/A$.

$X_1\underline{X_2}X_3A \xrightarrow{C} B \vdash X_1\underline{X_2}X_3 \xrightarrow{C} B/A$.

$(/ \to)$: (A) $Y_1\underline{Y_2} \xrightarrow{C_1} A \quad XB\underline{Z_1}Z_2 \xrightarrow{C_2} D \vdash X(B/A)Y_1\underline{Y_2Z_1}Z_2 \xrightarrow{C_1 \circ C_2} D$, where Y_2, Z_1 must be non-empty sequences.

$Y_1\underline{Y_2} \xrightarrow{C_1} A \quad X_1\underline{X_2B}Z \xrightarrow{C_2} C \vdash X_1\underline{X_2(B/A)Y_1}Y_2Z \xrightarrow{C_2/C_1} C$. If Y_2 is empty we may choose C_2 instead of C_2/C_1.

$Y \to A \quad X_1\underline{X_2BZ_1}Z_2 \xrightarrow{C} D \vdash X_1\underline{X_2(B/A)YZ_1}Z_2 \xrightarrow{C} D$. In the remaining cases copy the selection from the conclusion and the interpolant from the premise in which the selection is carried out.

$(\to \backslash)$ resp. $(\backslash \to)$: analogous to $(\to /)$ resp. $(/ \to)$.

$(\to \circ)$: $\underline{X} \xrightarrow{C_1} A \quad \underline{Y} \xrightarrow{C_2} B \vdash \underline{XY} \xrightarrow{C_1 \circ C_2} A \circ B$.

$X_1\underline{X_2} \stackrel{C_1}{\to} A \quad \underline{Y_1}Y_2 \stackrel{C_2}{\to} B \vdash X_1\underline{X_2Y_1}Y_2 \stackrel{C_1 \circ C_2}{\to} A \circ B$, where X_2, Y_1 must be non-empty. In the remaining cases again copy the selection from the conclusion and the interpolant from the premise in which the selection is carried out.

$(\circ \to)$: Copy the selection from the conclusion and the interpolant from the premise.

$(\to \wedge) : X \stackrel{C_1}{\to} A \quad X \stackrel{C_2}{\to} B \vdash X \stackrel{C_1 \wedge C_2}{\to} A \wedge B.$

$X_1\underline{X_2}X_3 \stackrel{C_1}{\to} A \quad X_1\underline{X_2}X_3 \stackrel{C_2}{\to} B \vdash X_1\underline{X_2}X_3 \stackrel{C_1 \wedge C_2}{\to} A \wedge B.$

$(\wedge \to) : X\underline{A}Y \stackrel{C}{\to} D \vdash X\underline{A \wedge B}Y \stackrel{C}{\to} D.$

$X\underline{B}Y \stackrel{C}{\to} D \vdash X\underline{A \wedge B}Y \stackrel{C}{\to} D$. In the remaining cases select in the premise what has also been selected in the conclusion (including one of the conjuncts) and take the interpolant from the premise.

$(\to \vee) : \underline{X} \stackrel{C}{\to} A \vdash \underline{X} \stackrel{C}{\to} A \vee B.$

$\underline{X} \stackrel{C}{\to} B \vdash \underline{X} \stackrel{C}{\to} A \vee B$. In the remaining cases copy the selection from the conclusion and the interpolant from the premise.

$(\vee \to)$: Select in the premises what has also been selected in the conclusion (including one disjunct). If C_1, C_2 are the interpolants provided by the premises, then $C_1 \vee C_2$ is the interpolant for the conclusion.

P, **C′** : Choose the interpolant from the premise.

C, **M** : $X\underline{AA}Y \stackrel{C}{\to} B \vdash X\underline{A}Y \stackrel{C}{\to} B.$

$X\underline{<>}Y \stackrel{\top}{\to} B \vdash X\underline{A}Y \stackrel{\top}{\to} B$. In the remaining cases select in the premise what has also been selected in the conclusion and take the interpolant from the premise. □

From the proof of the previous theorem we can derive as a corollary interpolation results for elementary fragments, i.e. sets of L-formulas which contain $PROP \cup \{\bot, \mathbf{t}, \top\}$ and which are closed under the connectives in a certain subset of $\{/, \backslash, \circ, \wedge, \vee\}$. The problem that may arise with fragments is to find interpolants which do not lead outside the fragment in question. Note e.g. the use of $\circ$ in the above induction step for $(/ \to)$. Indeed, for certain combinations of connectives interpolation may fail, as shown by Roorda [1991]. Note that the case (A) in the prove of the interpolation theorem does not arise if IND is not used for $(/ \to)$. The rules **P** and **C′** have not been subjected to IND because of selections like $XABY \to D \vdash \underline{XB}AY \to D$, $XAYAZ \to B \vdash XA\underline{YZ} \to B$.

Corollary 3.21 (i) Interpolation holds for all elementary fragments of Ξ_{Θ}.
(ii) Interpolation in the sense of IND holds for the elementary fragments of Ξ and $\Xi_{\{\mathbf{M}\}}$ based on: (a) $\{/, \circ\}$, $\{\backslash, \circ\}$, and $\{/, \backslash, \circ\}$, (b) every subset of $\{\circ, \wedge, \vee\}$, and the fragments obtained by joining any of the latter bases with one from (a).

Consider the following elementary fragments of Ξ and $\Xi_{\{\mathbf{P}\}}$: $(/, \backslash, \circ)$, $(/, \circ)$, $(\backslash, \circ)$. For each of them interpolation in the sense of IND can be strengthened (cf. [Roorda 1991]) by requiring that for every propositional variable p there are injections from the positive resp. negative occurrences of p in the interpolant C to those in X and those in A. Obviously, the rules $(\to \wedge)$, $(\vee \to)$ spoil this stronger interpolation property.

Chapter 4

Functional completeness for substructural subsystems of *IPL*

The problem of functional completeness for a given logic $\mathcal{L}$ is the problem of finding a set Γ of logical operations of $\mathcal{L}$ such that every logical operation of $\mathcal{L}$ is ***explicitly definable*** by a finite number of compositions from the elements of Γ. In this chapter, which is based on [Wansing 1990], we present a generalization of von Kutschera's [1968] approach to the problem of functional completeness for *IPL*. Besides Lorenz's [1968] analysis wrt a game-theoretical semantics for *IPL*, von Kutschera's proof seems to be the earliest published result on functional completeness for *IPL*. It makes use of a proof-theoretic interpretation specifying general rule schemata in a higher-level Gentzen-style sequent calculus and shows the set of intuitionistic connectives $\Gamma_1 = \{\neg, \wedge, \vee, \supset\}$ to be functionally complete for *IPL*. A proof of functional completeness of Γ_1 and $\Gamma_2 = \{\bot, \wedge, \vee, \supset\}$ wrt Kripke's semantics for *IPL* can be found in [McCullough 1971]. Inspired by [Kutschera 1968], Schroeder-Heister [1984] has proved functional completeness of Γ_2 for *IPL* wrt an extended natural deduction framework that allows for assumptions of arbitrary finite level. Functional completeness of Γ_2 for *IPL* wrt natural deduction has also been shown by Prawitz [1979]. Zucker & Tragesser [1978] consider a so-called 'inferential' interpretation of Gentzen's natural deduction according to which the meaning of each logical operation is given by its set of introduction rules. They show that in a natural deduction framework for every connective F one can find a finite combination of connectives from Γ_2 with the same set of introduction rules and thus with the same meaning as F. It is not clear, however, whether the presence of shared introduction rules implies exchangeability in all deductive contexts, which follows by explicit definability.

Up to now the problem of functional completeness for substructural subsystems of *IPL* has been open. Approaches like those of Prawitz and Schroeder-Heister which are based on Gentzen's natural deduction turn out not to be appropriate for cases where the order of premise occurrences matters, i.e. where in the framework of sequent calculi the structural rule **P** is missing. In the absence of **P** there is e.g. no natural deduction elimination rule capturing the rule for introducing $\circ$ on the lhs of the sequent arrow. (In sequent-style natural deduction there *is* an obvious elimination rule for $\circ$.) Thus, although Schroeder-Heister's approach to functional completeness for *IPL* can be (re)translated into a higher-level sequent calculus framework (see [Avron 1990]), the converse is not possible for subsystems of *IPL* which lack permutation. The aim of this chapter is to show how von Kutschera's approach to functional completeness for *IPL* can be applied to the substructural subsystems of *IPL* and *MPL* introduced in Chapter 3.

4.1 The higher-level Gentzen calculus G

Let L be any formal language and let $FORM(\mathsf{L})$ be the set of all L-formulas.

Definition 4.1 The set of all SL-formulas is the smallest set Γ such that

$FORM(\mathsf{L}) \subseteq \Gamma$;

$\bot \in \Gamma$;

if $T \in \Gamma$, then $(\rightarrow T)$, $(T \leftarrow) \in \Gamma$;

if $T_1, \ldots, T_n \in \Gamma$, then $(T_1 \ldots T_n \rightarrow)$, $(\leftarrow T_1 \ldots T_n) \in \Gamma$;

if $T_1, \ldots, T_n, U \in \Gamma$, then $(T_1 \ldots T_n \rightarrow U)$, $(U \leftarrow T_1 \ldots T_n) \in \Gamma$.

We shall use $T, U, T_1, T_2, \ldots$ as metavariables for SL-formulas and $\mathsf{X}, \mathsf{Y}, \mathsf{X}_1, \mathsf{X}_2, \ldots$ as metavariables for finite, possibly empty sequences of SL-formula occurrences. $\mathsf{Z}, \mathsf{Z}_1, \ldots$ will denote sequences of SL-formula occurrences with at most one element. Sometimes outermost parentheses in SL-formulas will be omitted.

Definition 4.2 Every $A \in FORM(\mathsf{L})$ is an SL-formula of S-degree 0;

$\bot$ is of S-degree 0;

if n is the maximum of the S-degrees of the SL-formulas in X, Z, then $\mathsf{X} \rightarrow \mathsf{Z}$, $\mathsf{Z} \leftarrow \mathsf{X}$ are called SL-formulas of S-degree $n + 1$.

If the S-degree of T is 1, then T is called a sequent. If the S-degree of $T = n$, we write $Sd(T) = n$. If $Sd(T) > 1$, then T is called a higher-level sequent.

Definition 4.3 Every SL-formula is an S-subformula of itself;

every S-subformula of an SL-formula in X, Z is an S-subformula of $\mathsf{X} \rightarrow \mathsf{Z}, \mathsf{Z} \leftarrow \mathsf{X}$.

The S-subformulas of T of S-degree 0 are called formula components of T. Let $T_1 \ldots T_n \Rightarrow \mathsf{Z}$ resp. $\Rightarrow \mathsf{Z}$ abbreviate $T_1 \ldots T_n \rightarrow \mathsf{Z}$ resp. $\rightarrow \mathsf{Z}$ as well as $\mathsf{Z} \leftarrow T_1 \ldots T_n$ resp. $\mathsf{Z} \leftarrow$. $\mathsf{X} \rightarrow$ resp. $\leftarrow \mathsf{X}$ is considered to be synonymous with $\mathsf{X} \rightarrow \bot$ resp. $\bot \leftarrow \mathsf{X}$. T is called the succedent of $\mathsf{X} \Rightarrow T$. An SL-formula T is called positive, if every S-subformula of T has a succedent.

Next, we shall define a basic, higher-level sequent calculus **G**, which like *MSPL* and *ISPL* but unlike von Kutschera's [1968] higher-level sequent calculus $\overline{K}_\infty$ is void of any structural rules of inference.

Definition 4.4 The rules of **G** are:

(ref)	$\vdash T \Rightarrow T$;
(tra)	$(\mathsf{X} \rightarrow T)(\mathsf{Y}_1 T \mathsf{Y}_2 \rightarrow T_1) \vdash \mathsf{Y}_1\mathsf{X}\mathsf{Y}_2 \rightarrow T_1$,
	$(T \leftarrow \mathsf{X})(T_1 \leftarrow \mathsf{Y}_1 T \mathsf{Y}_2) \vdash T_1 \leftarrow \mathsf{Y}_1\mathsf{X}\mathsf{Y}_2$;
$(\bot \Rightarrow)$	$\vdash \mathsf{X}\bot\mathsf{Y} \Rightarrow T$;
$(\rightarrow\rightarrow)$	$\mathsf{XY} \rightarrow T \vdash \mathsf{Y} \rightarrow (\mathsf{X} \rightarrow T)$;
$(\rightarrow\leftarrow)$	$\mathsf{YX} \rightarrow T \vdash \mathsf{Y} \rightarrow (T \leftarrow \mathsf{X})$;
$(\rightarrow\rightarrow)^-$	$\mathsf{Y} \rightarrow (\mathsf{X} \rightarrow T) \vdash \mathsf{XY} \rightarrow T$;
$(\rightarrow\leftarrow)^-$	$\mathsf{Y} \rightarrow (T \leftarrow \mathsf{X}) \vdash \mathsf{YX} \rightarrow T$;
$(\leftarrow\leftarrow)$	$T \leftarrow \mathsf{YX} \vdash (T \leftarrow \mathsf{X}) \leftarrow \mathsf{Y}$;
$(\leftarrow\rightarrow)$	$T \leftarrow \mathsf{XY} \vdash (\mathsf{X} \rightarrow T) \leftarrow \mathsf{Y}$;
$(\leftarrow\leftarrow)^-$	$(T \leftarrow \mathsf{X}) \leftarrow \mathsf{Y} \vdash T \leftarrow \mathsf{YX}$;
$(\leftarrow\rightarrow)^-$	$(\mathsf{X} \rightarrow T) \leftarrow \mathsf{Y} \vdash T \leftarrow \mathsf{XY}$.

Note that the rules $(\rightarrow\rightarrow)$ - $(\leftarrow\rightarrow)^-$ parallel the rules $(\rightarrow /)$, $(\rightarrow \backslash)$, $(\uparrow /)$, and $(\uparrow \backslash)$. The rules (ref) and (tra), of course, are higher-level counterparts of (id) and (cut), respectively.[1] $\mathcal{D}_{\mathbf{G}}(\Pi, T, \mathsf{X})$, "$\Pi$ is a derivation in **G** of T from the finite, possibly empty sequence X of SL-formula occurrences", is defined in a way induced by the rules of **G** (cf. Appendix 1.5).

Let U_V be an SL-formula which contains a certain occurrence of V as an S-subformula, and let U_T be the result of replacing this occurrence of V in U by T. Let moreover $V \Leftrightarrow T$ denote $V \Rightarrow T$ and $T \Rightarrow V$.

Theorem 4.5 If $\vdash V \Leftrightarrow T$ in **G**, then $\vdash U_V \Leftrightarrow U_T$ in **G**.

PROOF The proof is by induction on $n = Sd(U_V) - Sd(V)$. If $n = 0$, then $U_V = V$ and the claim is trivial. Suppose that the claim holds for every $n \leq m$, and $n = m+1$. Then U_V has the form $\mathsf{X}W_V\mathsf{Y} \rightarrow T_1$, $T_1 \leftarrow \mathsf{X}W_V\mathsf{Y}$, $\mathsf{X} \rightarrow W_V$, or $W_V \leftarrow \mathsf{X}$, where W_V contains the occurrence of V in question and $Sd(W_V) \leq n$. Now, suppose that $\vdash V \Leftrightarrow T$. By the induction hypothesis, $\vdash W_V \Leftrightarrow W_T$. Applying (tra) we obtain

$$\vdash (\mathsf{X}W_V\mathsf{Y} \rightarrow T_1) \Rightarrow (\mathsf{X}W_T\mathsf{Y} \rightarrow T_1), \quad \vdash (T_1 \leftarrow \mathsf{X}W_T\mathsf{Y}) \Rightarrow (T_1 \leftarrow \mathsf{X}W_V\mathsf{Y}),$$
$$\vdash (\mathsf{X}W_T\mathsf{Y} \rightarrow T_1) \Rightarrow (\mathsf{X}W_V\mathsf{Y} \rightarrow T_1), \quad \vdash (T_1 \leftarrow \mathsf{X}W_V\mathsf{Y}) \Rightarrow (T_1 \leftarrow \mathsf{X}W_T\mathsf{Y}),$$
$$\vdash (\mathsf{X} \rightarrow W_V) \Rightarrow (\mathsf{X} \rightarrow W_T), \quad \vdash (W_T \leftarrow \mathsf{X}) \Rightarrow (W_V \leftarrow \mathsf{X}),$$
$$\vdash (\mathsf{X} \rightarrow W_T) \Rightarrow (\mathsf{X} \rightarrow W_V), \quad \vdash (W_V \leftarrow \mathsf{X}) \Rightarrow (W_T \leftarrow \mathsf{X}).$$

Thus, $\vdash U_V \Leftrightarrow U_T$. □

[1] Since von Kutschera [1968] assumes permutabilty of premises, he considers only one sequent arrow, viz. $\rightarrow$.

4.2 Gentzen semantics

The higher-level sequent calculus G will now serve as a basis for introducing a proof-theoretic interpretation of propositional connectives. The basic idea of this interpretation is that an n-ary ($0 \leq n < \omega$) propositional connective F in the language L of a given logic $\mathcal{L}$ is characterized by the inference rules for introducing formulas $F(A_1, \ldots, A_n)$ into premises and conclusions. In order to be viewed as providing a genuine semantics, general rule schemata for introducing connectives into premises and conclusions are subject to the following constraints, analogous to those in [Kutschera 1968, p. 11]:

(i) Rule schemata characterizing F mention apart from one occurrence of F no other occurrence of a propositional connective; the role of formulas $F(A_1, \ldots, A_n)$ in deductive contexts depends on the deductive relationships between $A_1, \ldots, A_n$ only.

(ii) The rule schemata for F are non-creative ones, i.e. every proof of an F-free formula A in the result of extending G by instantiations of these schemata can be converted into a proof of A with no applications of rules characterizing F.

Constraint (i) suggests the following schemata for rules introducing $F(A_1, \ldots, A_n)$ on the rhs of $\rightarrow$ or on the lhs of $\leftarrow$:

$$(I)\,(a)\quad \mathsf{X}_{11}\mathsf{W}_{11}\mathsf{Y}_{11} \rightarrow \mathsf{Z}_{11} \;\ldots\; \mathsf{X}_{1s_1}\mathsf{W}_{1s_1}\mathsf{Y}_{1s_1} \rightarrow \mathsf{Z}_{1s_1} \vdash \mathsf{W}_{11}\ldots\mathsf{W}_{1s_1} \rightarrow F(A_1, \ldots, A_n),$$

$$\vdots$$

$$\mathsf{X}_{t1}\mathsf{W}_{t1}\mathsf{Y}_{t1} \rightarrow \mathsf{Z}_{t1} \;\ldots\; \mathsf{X}_{ts_t}\mathsf{W}_{ts_t}\mathsf{Y}_{ts_t} \rightarrow \mathsf{Z}_{ts_t} \vdash \mathsf{W}_{t1}\ldots\mathsf{W}_{ts_t} \rightarrow F(A_1, \ldots, A_n),$$
$$\mathsf{Z}_{11} \leftarrow \mathsf{X}_{11}\mathsf{W}_{11}\mathsf{Y}_{11} \;\ldots\; \mathsf{Z}_{1s_1} \leftarrow \mathsf{X}_{1s_1}\mathsf{W}_{1s_1}\mathsf{Y}_{1s_1} \vdash F(A_1, \ldots, A_n) \leftarrow \mathsf{W}_{11}\ldots\mathsf{W}_{1s_1},$$

$$\vdots$$

$$\mathsf{Z}_{t1} \leftarrow \mathsf{X}_{t1}\mathsf{W}_{t1}\mathsf{Y}_{t1} \;\ldots\; \mathsf{Z}_{ts_t} \leftarrow \mathsf{X}_{ts_t}\mathsf{W}_{ts_t}\mathsf{Y}_{ts_t} \vdash F(A_1, \ldots, A_n) \leftarrow \mathsf{W}_{t1}\ldots\mathsf{W}_{ts_t};$$

$$(I)\,(b)\quad \mathsf{X}_1\mathsf{W}\mathsf{Y}_1 \rightarrow \mathsf{Z}_1 \ldots \mathsf{X}_j\mathsf{W}\mathsf{Y}_j \rightarrow \mathsf{Z}_j \vdash \mathsf{W} \rightarrow F(A_1, \ldots, A_n),$$
$$\mathsf{Z}_1 \leftarrow \mathsf{X}_1\mathsf{W}\mathsf{Y}_1 \ldots \mathsf{Z}_j \leftarrow \mathsf{X}_j\mathsf{W}\mathsf{Y}_j \vdash F(A_1, \ldots, A_n) \leftarrow \mathsf{W}.$$

Here the W_{ik_i} ($i = 1, \ldots, t$; $k_i = 1, \ldots, s_i$) and W are unspecified sequences of SL-formula occurrences, whereas $\mathsf{X}_{ik_i}, \mathsf{Y}_{ik_i}, \mathsf{Z}_{ik_i}$ resp. X_1, ..., X_j, Y_1, ..., Y_j, Z_1, ..., Z_j contain only formula components from $A_1, \ldots, A_n$. Moreover, in each instantiation of (I) (a) resp. (I) (b) every A_k ($k = 1, \ldots, n$) occurs in some X_{ik_i}, Y_{ik_i}, or Z_{ik_i} resp. in some X_l, Y_l, or Z_l ($l = 1, \ldots, j$). If $n = 0$, then (I) (a) is $\vdash \Rightarrow F$, and (I) (b) is $\vdash \mathsf{W} \Rightarrow F$, where W is an unspecified sequence of SL-formula occurrences.

The schemata (I) (a) and (I) (b) are equivalent to the schemata

$$(I)'\,(a)\quad \mathsf{W}_{i1} \rightarrow (\mathsf{X}_{i1} \rightarrow (\mathsf{Z}_{i1} \leftarrow \mathsf{Y}_{i1})) \;\ldots\; \mathsf{W}_{is_i} \rightarrow (\mathsf{X}_{is_i} \rightarrow (\mathsf{Z}_{is_i} \leftarrow \mathsf{Y}_{is_i})) \vdash$$
$$\vdash \mathsf{W}_{i1}\ldots\mathsf{W}_{is_i} \rightarrow F(A_1, \ldots, A_n),$$
$$(\mathsf{X}_{i1} \rightarrow (\mathsf{Z}_{i1} \leftarrow \mathsf{Y}_{i1})) \leftarrow \mathsf{W}_{i1} \;\ldots\; (\mathsf{X}_{is_i} \rightarrow (\mathsf{Z}_{is_i} \leftarrow \mathsf{Y}_{is_i})) \leftarrow \mathsf{W}_{is_i} \vdash$$
$$\vdash F(A_1, \ldots, A_n) \leftarrow \mathsf{W}_{i1}\ldots\mathsf{W}_{is_i}, \quad i = 1, \ldots, t;$$

$$(I)'\,(b)\quad \mathsf{W} \rightarrow (\mathsf{X}_1 \rightarrow (\mathsf{Z}_1 \leftarrow \mathsf{Y}_1)) \;\ldots\; \mathsf{W} \rightarrow (\mathsf{X}_j \rightarrow (\mathsf{Z}_j \leftarrow \mathsf{Y}_j)) \vdash$$
$$\vdash \mathsf{W} \rightarrow F(A_1, \ldots, A_n),$$
$$(\mathsf{X}_1 \rightarrow (\mathsf{Z}_1 \leftarrow \mathsf{Y}_1)) \leftarrow \mathsf{W} \ldots (\mathsf{X}_j \rightarrow (\mathsf{Z}_j \leftarrow \mathsf{Y}_j)) \leftarrow \mathsf{W} \vdash$$
$$\vdash F(A_1, \ldots, A_n) \leftarrow \mathsf{W},$$

where if $n = 0$, $(I)'$ (a) is $\vdash \Rightarrow F$ and $(I)'$ (b) is $\mathsf{W} \Rightarrow F$. Since premise occurrences are removed only by applications of the higher-level (cut)-rule (tra), constraint (ii) amounts to the requirement that applications of (tra) with a cut-formula $F(A_1, \ldots, A_n)$ can be eliminated. [2] To ensure this, the schemata for rules introducing $F(A_1, \ldots, A_n)$ into premises become:

$$(II)(a)\quad \mathsf{Y}_1\mathsf{X}_1\mathsf{Y}_2 \to \mathsf{Z} \ \ldots\ \mathsf{Y}_1\mathsf{X}_t\mathsf{Y}_2 \to \mathsf{Z} \vdash \mathsf{Y}_1F(A_1, \ldots, A_n)\mathsf{Y}_2 \to \mathsf{Z},$$
$$\mathsf{Z} \leftarrow \mathsf{Y}_1\mathsf{X}_1\mathsf{Y}_2 \ \ldots\ \mathsf{Z} \leftarrow \mathsf{Y}_1\mathsf{X}_t\mathsf{Y}_2 \vdash \mathsf{Z} \leftarrow \mathsf{Y}_1F(A_1, \ldots, A_n)\mathsf{Y}_2;$$
$$(II)\,(b)\quad \mathsf{Y}_1(\mathsf{X}_l \to (\mathsf{Z}_l \leftarrow \mathsf{Y}_l))\mathsf{Y}_2 \to \mathsf{Z} \vdash \mathsf{Y}_1F(A_1, \ldots, A_n)\mathsf{Y}_2 \to \mathsf{Z},$$
$$\mathsf{Z} \leftarrow \mathsf{Y}_1(\mathsf{X}_l \to (\mathsf{Z}_l \leftarrow \mathsf{Y}_l))\mathsf{Y}_2 \vdash \mathsf{Z} \leftarrow \mathsf{Y}_1F(A_1, \ldots, A_n)\mathsf{Y}_2, \quad l = 1, \ldots, j;$$

where $\mathsf{Y}_1, \mathsf{Y}_2$ are unspecified sequences of $S\mathsf{L}$-formula occurrences and $\mathsf{X}_i = \mathsf{X}_{i1} \to (\mathsf{Z}_{i1} \leftarrow \mathsf{Y}_{i1}) \ldots \mathsf{X}_{is_i} \to (\mathsf{Z}_{is_i} \leftarrow \mathsf{Y}_{is_i})$. If $n = 0$, then (II) (a) is $\mathsf{Y}_1\mathsf{Y}_2 \to \mathsf{Z} \vdash \mathsf{Y}_1F\mathsf{Y}_2 \to \mathsf{Z}$, $\mathsf{Z} \leftarrow \mathsf{Y}_1\mathsf{Y}_2 \vdash \mathsf{Z} \leftarrow \mathsf{Y}_1F\mathsf{Y}_2$, and (II) (b) is not instantiated.

Thus, the rule schemata (I) (a) resp. (I) (b) for introductions into conclusions already completely determine the schemata (II) (a) resp. (II) (b) for introductions into premises. The rules of $\mathbf{G}$ $+(I)$+(II) determine the ways in which forumlas $F(A_1, \ldots, A_n)$ may be introduced into arbitrary $S\mathsf{L}$-formula contexts.

The schemata (I) (a) apparently impose a certain restriction on the Gentzen semantics: 'infix-operations' are to be excluded. However, allowing for infix-operations (i) blocks cut-elimination and (ii) leads outside the present semantical framework; e.g. with the following binary infix-operation $\bowtie$:

$$(\Rightarrow\bowtie)\quad \mathsf{X}T\mathsf{Y} \to U \vdash \mathsf{X}\mathsf{Y} \to (T \bowtie U),$$
$$U \leftarrow \mathsf{X}T\mathsf{Y} \vdash (T \bowtie U) \leftarrow \mathsf{X}\mathsf{Y};$$
$$(\bowtie\Rightarrow)\quad \mathsf{X}(T \to U)\mathsf{Y} \to V \vdash \mathsf{X}(T \bowtie U)\mathsf{Y} \to V,\quad \mathsf{X}(T \leftarrow U)\mathsf{Y} \to V \vdash \mathsf{X}(T \bowtie U)\mathsf{Y} \to V,$$
$$V \leftarrow \mathsf{X}(T \to U)\mathsf{Y} \vdash V \leftarrow \mathsf{X}(T \bowtie U)\mathsf{Y},\quad V \leftarrow \mathsf{X}(T \leftarrow U)\mathsf{Y} \vdash V \leftarrow \mathsf{X}(T \bowtie U)\mathsf{Y};$$

permutation of premise occurrences becomes derivable, as can easily be verified. Thus, the addition of $\bowtie$ to $\mathbf{G}$ $+(I) + (II)$ is *not* conservative.

Let C_A denote an L-formula which contains a certain occurrence of A as a subformula, and let C_B denote the result of replacing this occurrence of A in C by B. The degree of A ($d(A)$) is the number of occurrences of propositional connectives in A. If $\mathsf{X} = T_1 \ldots T_m$, then X_A denotes $T_{1A} \ldots T_{mA}$.

Theorem 4.6 If $\vdash A \Leftrightarrow B$ in $\mathbf{G}$ $+(I) + (II)$, then $\vdash C_A \Leftrightarrow C_B$ in $\mathbf{G}$ $+(I) + (II)$.

PROOF By induction on $l = d(C_A) - d(A)$. If $l = 0$, the proof is trivial. Suppose that the claim holds for every $l \leq m$, and $l = m + 1$. Then C_A has the form $F(A_1, \ldots, A_n)$, where one of the A_{kA} contains the occurrence of A in question and $d(A_{kA}) \leq l$. Suppose that $\vdash A \Leftrightarrow B$. By the induction hypothesis, $\vdash A_{kA} \Leftrightarrow A_{kB}$, and by Theorem 4.5, $\vdash (\mathsf{X}_{is_i} \to (\mathsf{Z}_{is_i} \leftarrow \mathsf{Y}_{is_i}))_A \Leftrightarrow (\mathsf{X}_{is_i} \to (\mathsf{Z}_{is_i} \leftarrow \mathsf{Y}_{is_i}))_B$, $\vdash (\mathsf{X}_l \to (\mathsf{Z}_l \leftarrow \mathsf{Y}_l))_A \Leftrightarrow (\mathsf{X}_l \to (\mathsf{Z}_l$

[2] Note that von Kutschera's higher-level version of (cut) takes advantage of the structural inference rules assumed.

$\leftarrow \mathsf{Y}_l))_B$, where in each case the replacement of A by B is wrt A_k. Suppose the rules for C_A are instantiations of $(I)\,(a)$ and $(II)\,(a)$. By (ref) and the schemata $(I)\,(a)$ we obtain $\vdash \mathsf{X}_{i_A} \rightarrow C_B$, $\vdash \mathsf{X}_{i_B} \rightarrow C_A$, $\vdash C_B \leftarrow \mathsf{X}_{i_A}$, $\vdash C_A \leftarrow \mathsf{X}_{i_B}$. The schemata $(II)\,(a)$ and (tra) give

$$\mathsf{X}_{1_A} \rightarrow C_B \ldots \mathsf{X}_{t_A} \rightarrow C_B \vdash C_A \rightarrow C_B, \quad \mathsf{X}_{1_B} \rightarrow C_A \ldots \mathsf{X}_{t_B} \rightarrow C_A \vdash C_B \rightarrow C_A,$$

$$C_B \leftarrow \mathsf{X}_{1_A} \ldots C_B \leftarrow \mathsf{X}_{t_A} \vdash C_B \leftarrow C_A, \quad C_A \leftarrow \mathsf{X}_{1_B} \ldots C_A \leftarrow \mathsf{X}_{t_B} \vdash C_A \leftarrow C_B.$$

Thus, $\vdash C_A \Leftrightarrow C_B$. If the rules for C_A are instantiations of $(I)\,(b)$ and $(II)\,(b)$, then by the induction hypothesis and Theorem 4.5, the schemata $(II)\,(b)$ give $\vdash C_A \Rightarrow (\mathsf{X}_l \rightarrow (\mathsf{Z}_l \leftarrow \mathsf{Y}_l))_B$ and $\vdash C_B \Rightarrow (\mathsf{X}_l \rightarrow (\mathsf{Z}_l \leftarrow \mathsf{Y}_l))_A$. By $(I)\,(b)$, we obtain $\vdash C_A \Leftrightarrow C_B$. □

Let T_A denote an SL-formula which contains a certain occurrence of A as a subformula of a formula component of T.

Theorem 4.7 If $\vdash A \Leftrightarrow B$ in **G** $+(I)+(II)$, then $\vdash T_A \Leftrightarrow T_B$ in **G** $+(I)+(II)$.

PROOF By the previous two theorems. □

4.3 Functional completeness for *ISPL*

In a first step we shall show that $\{/, \backslash, \wedge, \circ, \vee, \top, \bot\}$ is functionally complete wrt the Gentzen semantics. We shall define (i) rules for introducing the propositional constant **t** into conclusions and (ii) rules for introducing the binary connectives $/, \backslash, \wedge, \circ, \vee$ and the constant $\top$ into premises and conclusions. These rules conform to the schemata $(I)\,(a)$ and $(II)\,(a)$ resp. $(I)\,(b)$ and $(II)\,(b)$:

$\underline{(\Rightarrow \mathbf{t})}$ $\vdash \mathsf{X} \Rightarrow \mathbf{t}$;

$\underline{(\Rightarrow /)'}$ $\mathsf{X} \rightarrow (U \leftarrow T) \vdash \mathsf{X} \rightarrow (U/T)$,
$(U \leftarrow T) \leftarrow \mathsf{X} \vdash (U/T) \leftarrow \mathsf{X}$;

$\underline{(/ \Rightarrow)'}$ $\mathsf{Y}_1(U \leftarrow T)\mathsf{Y}_2 \rightarrow \mathsf{Z} \vdash \mathsf{Y}_1(U/T)\mathsf{Y}_2 \rightarrow \mathsf{Z}$,
$\mathsf{Z} \leftarrow \mathsf{Y}_1(U \leftarrow T)\mathsf{Y}_2 \vdash \mathsf{Z} \leftarrow \mathsf{Y}_1(U/T)\mathsf{Y}_2$;

$\underline{(\Rightarrow \backslash)'}$ $\mathsf{X} \rightarrow (T \rightarrow U) \vdash \mathsf{X} \rightarrow (T \backslash U)$,
$(T \rightarrow U) \leftarrow \mathsf{X} \vdash (T \backslash U) \leftarrow \mathsf{X}$;

$\underline{(\backslash \Rightarrow)'}$ $\mathsf{Y}_1(T \rightarrow U)\mathsf{Y}_2 \rightarrow \mathsf{Z} \vdash \mathsf{Y}_1(T \backslash U)\mathsf{Y}_2 \rightarrow \mathsf{Z}$,
$\mathsf{Z} \leftarrow \mathsf{Y}_1(T \rightarrow U)\mathsf{Y}_2 \vdash \mathsf{Z} \leftarrow \mathsf{Y}_1(T \backslash U)\mathsf{Y}_2$;

$\underline{(\Rightarrow \wedge)}$ $(\mathsf{X} \rightarrow T)(\mathsf{X} \rightarrow U) \vdash \mathsf{X} \rightarrow (T \wedge U)$,
$(T \leftarrow \mathsf{X})(U \leftarrow \mathsf{X}) \vdash (T \wedge U) \leftarrow \mathsf{X}$;

$\underline{(\wedge \Rightarrow)}$ $\mathsf{Y}_1 T \mathsf{Y}_2 \rightarrow \mathsf{Z} \vdash \mathsf{Y}_1(T \wedge U)\mathsf{Y}_2 \rightarrow \mathsf{Z}$,
$\mathsf{Z} \leftarrow \mathsf{Y}_1 T \mathsf{Y}_2 \vdash \mathsf{Z} \leftarrow \mathsf{Y}_1(T \wedge U)\mathsf{Y}_2$,
$\mathsf{Y}_1 U \mathsf{Y}_2 \rightarrow \mathsf{Z} \vdash \mathsf{Y}_1(T \wedge U)\mathsf{Y}_2 \rightarrow \mathsf{Z}$,
$\mathsf{Z} \leftarrow \mathsf{Y}_1 U \mathsf{Y}_2 \vdash \mathsf{Z} \leftarrow \mathsf{Y}_1(T \wedge U)\mathsf{Y}_2$;

$\underline{(\Rightarrow \circ)}$ $(\mathsf{X} \rightarrow T)(\mathsf{Y} \rightarrow U) \vdash \mathsf{XY} \rightarrow (T \circ U)$,
$(T \leftarrow \mathsf{X})(U \leftarrow \mathsf{Y}) \vdash (T \circ U) \leftarrow \mathsf{XY}$;

$\underline{(\circ \Rightarrow)}$ $\mathsf{Y}_1 TU \mathsf{Y}_2 \rightarrow \mathsf{Z} \vdash \mathsf{Y}_1(T \circ U)\mathsf{Y}_2 \rightarrow \mathsf{Z}$,
$\mathsf{Z} \leftarrow \mathsf{Y}_1 TU \mathsf{Y}_2 \vdash \mathsf{Z} \leftarrow \mathsf{Y}_1(T \circ U)\mathsf{Y}_2$;

$\underline{(\Rightarrow \vee)}$ $\mathsf{X} \rightarrow T \vdash \mathsf{X} \rightarrow (T \vee U)$,
$\mathsf{X} \rightarrow U \vdash \mathsf{X} \rightarrow (T \vee U)$,
$T \leftarrow \mathsf{X} \vdash (T \vee U) \leftarrow \mathsf{X}$,
$U \leftarrow \mathsf{X} \vdash (T \vee U) \leftarrow \mathsf{X}$;

$\underline{(\vee \Rightarrow)}$ $(\mathsf{Y}_1 T \mathsf{Y}_2 \rightarrow \mathsf{Z})(\mathsf{Y}_1 U \mathsf{Y}_2 \rightarrow \mathsf{Z}) \vdash \mathsf{Y}_1(T \vee U)\mathsf{Y}_2 \rightarrow \mathsf{Z}$,
$(\mathsf{Z} \leftarrow \mathsf{Y}_1 T \mathsf{Y}_2)(\mathsf{Z} \leftarrow \mathsf{Y}_1 U \mathsf{Y}_2) \vdash \mathsf{Z} \leftarrow \mathsf{Y}_1(T \vee U)\mathsf{Y}_2$;

$\underline{(\Rightarrow \top)}$ $\vdash \Rightarrow \top$;

$\underline{(\top \Rightarrow)}$ $\mathsf{Y}_1\mathsf{Y}_2 \rightarrow \mathsf{Z} \vdash \mathsf{Y}_1 \top \mathsf{Y}_2 \rightarrow \mathsf{Z}$,
$\mathsf{Z} \leftarrow \mathsf{Y}_1\mathsf{Y}_2 \vdash \mathsf{Z} \leftarrow \mathsf{Y}_1 \top \mathsf{Y}_2$.

It can readily be seen that (i) $\underline{(\Rightarrow /)'}$ resp. $\underline{(\Rightarrow \backslash)'}$ is equivalent to

$\underline{(\Rightarrow /)}$ $\mathsf{X}T \rightarrow U \vdash \mathsf{X} \rightarrow (U/T)$,
$U \leftarrow \mathsf{X}T \vdash (U/T) \leftarrow \mathsf{X}$; resp.

$\underline{(\Rightarrow \backslash)}$ $T\mathsf{X} \rightarrow U \vdash \mathsf{X} \rightarrow (T \backslash U)$,
$U \leftarrow T\mathsf{X} \vdash (T \backslash U) \leftarrow \mathsf{X}$,

and (ii) $\underline{(/ \Rightarrow)'}$ resp. $\underline{(\backslash \Rightarrow)'}$ is equivalent to

$\underline{(/ \Rightarrow)}$ $\mathsf{X} \rightarrow (U/T) \vdash \mathsf{X}T \rightarrow U$,
$(U/T) \leftarrow \mathsf{X} \vdash U \leftarrow \mathsf{X}T$; resp.

$\underline{(\backslash \Rightarrow)}$ $\mathsf{X} \rightarrow (T \backslash U) \vdash T\mathsf{X} \rightarrow U$, $(T \backslash U) \leftarrow \mathsf{X} \vdash U \leftarrow T\mathsf{X}$.

Next, we assign to each $S\mathsf{L}$-formula T one formula $\overline{T}$ by stipulating:

if T is an L-formula, then $\overline{T} = T$;

if $\mathsf{X} = T_1 \ldots T_n$, then $\overline{\mathsf{X}} = \overline{T_1} \circ \ldots \circ \overline{T_n}$;

if $T = \mathsf{X} \rightarrow U$, then $\overline{T} = \overline{\mathsf{X}} \backslash \overline{U}$;

if $T = U \leftarrow \mathsf{X}$, then $\overline{T} = \overline{U}/\overline{\mathsf{X}}$;

if $T = \mathsf{X} \rightarrow$, then $\overline{T} = \overline{\mathsf{X}} \backslash \bot$;

if $T = \ \leftarrow \mathsf{X}$, then $\overline{T} = \bot/\overline{\mathsf{X}}$;

if $T = \ \rightarrow U$, then $\overline{T} = \top \backslash \overline{U}$;

if $T = U \leftarrow$, then $\overline{T} = \overline{U}/\top$;

if $T = \bot$, then $\overline{T} = \bot$.

Note that $Sd(\overline{T}) = 0$.

Theorem 4.8 In $\mathbf{G} + (I) + (II) \vdash T \Leftrightarrow \overline{T}$.

Proof By induction on $Sd(T)$. If $Sd(T) = 0$, the claim is trivial. Suppose that the claim holds for every $l \leq m$, and $l = m + 1$. By the induction hypothesis we have that $\vdash (T_1 \ldots T_n \rightarrow U) \Leftrightarrow (\overline{T_1} \ldots \overline{T_n} \rightarrow \overline{U})$ resp. $\vdash (U \leftarrow T_1 \ldots T_n) \Leftrightarrow (\overline{U} \leftarrow \overline{T_1} \ldots \overline{T_n})$ resp. $\vdash (T_1 \ldots T_n \rightarrow) \Leftrightarrow (\overline{T_1} \ldots \overline{T_n} \rightarrow)$ resp. $\vdash (\leftarrow T_1 \ldots T_n) \Leftrightarrow (\leftarrow \overline{T_1} \ldots \overline{T_n})$ resp. $\vdash (\rightarrow T) \Leftrightarrow (\rightarrow \overline{T})$ resp. $\vdash (T \leftarrow) \Leftrightarrow (\overline{T} \leftarrow)$. Now, by $\underline{(\Rightarrow \circ)}$, $\underline{(\circ \Rightarrow)}$, and (tra), we obtain $\vdash (\overline{T_1} \ldots \overline{T_n} \rightarrow \overline{U}) \Leftrightarrow (\overline{T_1} \circ \ldots \circ \overline{T_n} \rightarrow \overline{U})$; $\vdash (\overline{U} \leftarrow \overline{T_1} \ldots \overline{T_n}) \Leftrightarrow (\overline{U} \leftarrow \overline{T_1} \circ \ldots \circ \overline{T_n})$. By $\underline{(\Rightarrow /)}$ resp. $\underline{(\Rightarrow \backslash)}$ and $\underline{(/ \Rightarrow)}$ resp. $\underline{(\backslash \Rightarrow)}$ we obtain $\vdash (\overline{T} \rightarrow \overline{U}) \Leftrightarrow (\overline{T} \backslash \overline{U})$ resp. $\vdash (\overline{U} \leftarrow \overline{T}) \Leftrightarrow (\overline{U}/\overline{T})$. Finally, for $\overline{T} \rightarrow$ resp. $\leftarrow \overline{T}$, i.e. $\overline{T} \rightarrow \bot$ resp. $\bot \leftarrow \overline{T}$, we have $\vdash (\overline{T} \rightarrow \bot) \Leftrightarrow (\overline{T} \backslash \bot)$ resp. $\vdash (\bot \leftarrow \overline{T}) \Leftrightarrow (\bot/\overline{T})$; for $\rightarrow \overline{T}$ resp. $\overline{T} \leftarrow$ we have $\vdash (\rightarrow \overline{T}) \Leftrightarrow (\top \backslash \overline{T})$ resp. $\vdash (\overline{T} \leftarrow) \Leftrightarrow (\overline{T}/\top)$. These observations immediately establish the claim for $l = m + 1$. □

Theorem 4.9 $\{/, \backslash, \wedge, \circ, \vee, \top, \bot\}$ is functionally complete wrt the Gentzen semantics.

Proof Suppose that $F(A_1, \ldots, A_n)$ is defined by instantiations of the schemata (I) (a) and (II) (a). If $n = 0$, then $F = \top$. Otherwise, from (ref) and (I) (a) we obtain $\vdash \mathsf{X}_i \Rightarrow F(A_1, \ldots, A_n)$. By the previous theorem, (tra), and $\underline{(\circ \Rightarrow)}$, we obtain $\vdash \overline{\mathsf{X}_i} \Rightarrow F(A_1, \ldots, A_n)$. Finally, applications of $\underline{(\vee \Rightarrow)}$ give $\vdash \overline{\mathsf{X}_1} \vee \ldots \vee \overline{\mathsf{X}_t} \Rightarrow F(A_1, \ldots, A_n)$. By (ref), the previous theorem, (tra), and $\underline{(\Rightarrow \circ)}$, $\vdash \mathsf{X}_i \Rightarrow \overline{\mathsf{X}_i}$. Applying $\underline{(\Rightarrow \vee)}$, we obtain $\vdash \mathsf{X}_i \Rightarrow \overline{\mathsf{X}_1} \vee \ldots \vee \overline{\mathsf{X}_t}$. The schemata (II) (a) give $\vdash F(A_1, \ldots, A_n) \Rightarrow \overline{\mathsf{X}_1} \vee \ldots \vee \overline{\mathsf{X}_t}$. By Theorem 4.7, $F(A_1, \ldots, A_n)$ and $\overline{\mathsf{X}_1} \vee \ldots \vee \overline{\mathsf{X}_t}$ are interchangeable in $S\mathsf{L}$-formulas (wrt provable equivalence in terms of $\Leftrightarrow$). Thus, $F(A_1, \ldots, A_n)$ can be explicitly defined by a formula in $\{/, \backslash, \circ, \vee, \top, \bot\}$. Suppose now that the rules for $F(A_1, \ldots, A_n)$ are instantiations of (I) (b) and (II) (b). If $n = 0$, then $F = \mathbf{t}$, which is definable as $\bot/\bot$. Otherwise, by (ref), the previous theorem, and (tra), $\vdash \overline{S_l} \Rightarrow S_l$, where $S_l = \mathsf{X}_l \rightarrow (\mathsf{Z}_l \leftarrow \mathsf{Y}_l)$ $(1 \leq l \leq j)$. Repeatedly applying $\underline{(\wedge \Rightarrow)}$, we obtain $\vdash \overline{S_1} \wedge \ldots \wedge \overline{S_j} \Rightarrow S_l$. The schemata (I) (b) give $\vdash \overline{S_1} \wedge \ldots \wedge \overline{S_j} \Rightarrow F(A_1, \ldots, A_n)$. By (ref), the previous theorem, (tra), and $\underline{(\Rightarrow \circ)}$, we have $\vdash S_l \Rightarrow \overline{S_l}$. Using the schemata (II) (b) we may conclude that $\vdash F(A_1, \ldots, A_n) \Rightarrow \overline{S_l}$. Eventually, applications of $\underline{(\Rightarrow \wedge)}$ give $\vdash F(A_1, \ldots, A_n) \Rightarrow \overline{S_1} \wedge \ldots \wedge \overline{S_j}$. By Theorem 4.7, $F(A_1, \ldots, A_n)$ and $\overline{S_1} \wedge \ldots \wedge \overline{S_j}$ are intersubstitutable in $S\mathsf{L}$-formulas. Thus, $F(A_1, \ldots, A_n)$ can be explicitly defined by a formula in $\{/, \backslash, \circ, \wedge, \bot\}$. □

Von Kutschera's approach to the problem of functional completeness, as applied in the proof of the previous theorem, has (although to a certain extent independently of von Kutschera's work) become the standard proof-theoretic methodology (usually mutatis mutandis associated with systems of natural deduction): (i) the definiens can (more or less) be read off from the schemata for introductions into conclusions, making sure that the definiendum is derivable from the definiens, and (ii) the rules for introductions into premises can be used to show that conversely the definiens is derivable from the definiendum.

It remains to be shown that the Gentzen semantics characterizes *ISPL*.

Theorem 4.10 $\{/, \backslash, \wedge, \circ, \vee, \top, \bot\}$ is functionally complete for *ISPL*.

Proof If (i) SL-formulas and the premises and conclusions of $(\bot \Rightarrow)$, $(\Rightarrow \mathbf{t})$, $(\Rightarrow F)$, and $(F \Rightarrow)$ $(F \in \{\top, /, \backslash, \wedge, \circ, \vee\})$ are restricted to positive sequents only, and (ii) $T \leftarrow T_1 \ldots T_n$ is read as $T_1 \ldots T_n \rightarrow T$, then the resulting calculus is equivalent to *ISPL* in the sense that both systems have the same set of provable sequents, as a comparison of these systems immediately reveals. □

A natural question that might arise is why at all one should make use of the higher-level sequent calculus **G**, since after all $\vdash (B/A) \Leftrightarrow (B \leftarrow A)$ and $\vdash (A\backslash B) \Leftrightarrow (A \rightarrow B)$. Now, as has already been pointed out by von Kutschera for the case of *IPL*, one in fact *needs* the higher-level framework for proving functional completeness. Let $X' \Rightarrow X$ abbreviate $C'_j \Rightarrow C_j$, where $X' = C'_1 \ldots C'_m$, $X = C_1 \ldots C_m$, and $0 < j \leq m$. If there were no iterated sequent arrows, one would, according to the earlier constraints, in place of (I) (a) and (II) (a) obtain the schemata:

$$\begin{array}{ll}
(I)^\dagger & X_{11}W_{11}Y_{11} \rightarrow B_{11} \;\ldots\; X_{1s_1}W_{1s_1}Y_{1s_1} \rightarrow B_{1s_1} \vdash W_{11} \ldots W_{1s_1} \rightarrow F(A_1, \ldots, A_n), \\
& \vdots \\
& X_{t1}W_{t1}Y_{t1} \rightarrow B_{t1} \;\ldots\; X_{ts_t}W_{ts_t}Y_{ts_t} \rightarrow B_{ts_t} \vdash W_{t1} \ldots W_{ts_t} \rightarrow F(A_1, \ldots, A_n), \\
& B_{11} \leftarrow X_{11}W_{11}Y_{11} \;\ldots\; B_{1s_1} \leftarrow X_{1s_1}W_{1s_1}Y_{1s_1} \vdash F(A_1, \ldots, A_n) \leftarrow W_{11} \ldots W_{1s_1}, \\
& \vdots \\
& B_{t1} \leftarrow X_{t1}W_{t1}Y_{t1} \;\ldots\; B_{ts_t} \leftarrow X_{ts_t}W_{ts_t}Y_{ts_t} \vdash F(A_1, \ldots, A_n) \leftarrow W_{t1} \ldots W_{ts_t},
\end{array}$$

and

$$\begin{array}{ll}
(II)^\dagger & X'_{11} \rightarrow X_{11} \ldots X'_{ts_t} \rightarrow X_{ts_t} \;\; Y'_{11} \rightarrow Y_{11} \ldots Y'_{ts_t} \rightarrow Y_{ts_t} \;\; Z_1B_{11}Z_2 \rightarrow C \ldots \\
& \ldots Z_1B_{ts_t}Z_2 \rightarrow C \vdash Z_1X'_{11} \ldots X'_{ts_t}F(A_1, \ldots, A_n)Y'_{11} \ldots Y'_{ts_t}Z_2 \rightarrow C, \\
& X_{11} \leftarrow X'_{11} \ldots X_{ts_t} \leftarrow X'_{ts_t} \;\; Y_{11} \leftarrow Y'_{11} \ldots Y_{ts_t} \leftarrow Y'_{ts_t} \;\; C \leftarrow Z_1B_{11}Z_2 \ldots \\
& \ldots C \leftarrow Z_1B_{ts_t}Z_2 \vdash C \leftarrow Z_1X'_{11} \ldots X'_{ts_t}F(A_1, \ldots, A_n)Y'_{11} \ldots Y'_{ts_t}Z_2,
\end{array}$$

where the case $n = 0$ is treated as before, the W_{ik_i}, Y'_{ik_i}, and Z_1, Z_2 $(i = 1, \ldots, t;$ $k_i = 1, \ldots, s_i)$ are unspecified sequences of L-formula occurrences, every formula in X_{ik_i}, Y_{ik_i}, and each B_{ik_i} is among $A_1, \ldots, A_n$, and in each instantiation of $(I)^\dagger$ every A_k $(k = 1, \ldots, n)$ occurs in some X_{ik_i}, Y_{ik_i}, or B_{ik_i}. Now, although $\vdash \overline{\mathsf{X}_1} \vee \ldots \vee \overline{\mathsf{X}_t} \Rightarrow F(A_1, \ldots, A_n)$ (where $\mathsf{X}_i = X_{i1} \rightarrow (B_{i1} \leftarrow Y_{i1}) \ldots X_{is_i} \rightarrow (B_{is_i} \leftarrow Y_{is_i})$), we do not have $\vdash F(A_1, \ldots, A_n) \Rightarrow \overline{\mathsf{X}_1} \vee \ldots \vee \overline{\mathsf{X}_t}$. Consider e.g. the following connective F:

$$\begin{array}{ll}
(\rightarrow F) & AX \rightarrow B \vdash X \rightarrow F(A, B, C), \quad CX \rightarrow B \vdash X \rightarrow F(A, B, C); \\
(F \Rightarrow) & Y_1 \rightarrow A \quad Y_2 \rightarrow C \;\; Z_1BZ_2 \rightarrow D \vdash Z_1Y_1Y_2F(A, B, C)Z_2 \rightarrow D, \\
& A \leftarrow Y_1 \quad C \leftarrow Y_2 \;\; D \leftarrow Z_1BZ_2 \vdash D \leftarrow Z_1Y_1Y_2F(A, B, C)Z_2.
\end{array}$$

In the higher-level proof-theoretic semantics $F(A, B, C)$ would be explicitly definable by $(A\backslash B)\vee(C\backslash B)$ on the strength of $(\rightarrow F)$. Instead of $\vdash F(A, B, C) \Rightarrow (A\backslash B)\vee(C\backslash B)$ the semantics based on $(I)^\dagger$, $(II)^\dagger$ just gives $\vdash F(A, B, C) \Rightarrow (A \circ C) \backslash B$. However, $\nvdash_{ISPL} (A \circ C) \backslash B \rightarrow (A \backslash B) \vee (C \backslash B)$.

4.4 Functional completeness for $ISPL_\Delta$ and $MSPL_\Delta$

The present generalization of von Kutschera's approach to the problem of functional completeness for IPL extends directly to those subsystems of IPL which are obtained from $ISPL$ by adding some of the earlier-mentioned structural rules of inference. For every non-empty $\Delta \subseteq \{\mathbf{P}, \mathbf{C}, \mathbf{C}', \mathbf{E}, \mathbf{E}', \mathbf{M}\}$ one may add the higher-level formulation of the rules in Δ (using $\leftarrow$ as well as $\rightarrow$) to $\mathbf{G}$ and define the notion of derivation in analogy to the notion of derivation in $\mathbf{G}$. In the case of $ISPL_{\{\mathbf{C}\}}$ e.g. it has to be required that if $\mathcal{D}_{\mathbf{G}_{\{\mathbf{C}\}}}(\Pi, \mathsf{X}_1 UU \mathsf{X}_2 \rightarrow T, \mathsf{Y})$, then $\mathcal{D}_{\mathbf{G}_{\{\mathbf{C}\}}}(\frac{\Pi}{\mathsf{X}_1 U \mathsf{X}_2 \rightarrow T}, \mathsf{X}_1 U \mathsf{X}_2 \rightarrow T, \mathsf{Y})$ etcetera. The argument then is the same as for $ISPL$, i.e., in particular, the set of logical operations for which functional completeness is shown remains the same in each case:

Corollary 4.11 $\{/, \backslash, \wedge, \circ, \vee, \top, \bot\}$ is functionally complete for $ISPL_\Delta$.

As already pointed out, A/B and $B\backslash A$ are interderivable in the presence of $\mathbf{P}$, $A \wedge B$ and $A \circ B$ are interderivable in the presence of $\mathbf{M}$ and $\mathbf{C}$, and $\top$ and $\mathbf{t}$ are interderivable in the presence of $\mathbf{M}$.

Due to the absence of the *ex falso* principle $(\bot \;\rightarrow)$, $\mathbf{t}$ is no longer definable in $MSPL_\Delta$. Let $\mathbf{G}^m$ denote the result of dropping $(\bot \Rightarrow)$ from $\mathbf{G}$. Using $\mathbf{G}^m$ instead of $\mathbf{G}$ as the underlying proof-theoretic framework for introducing connectives into premises and conclusions, it is an immediate corollary to the above results that

Corollary 4.12 $\{/, \backslash, \wedge, \circ, \vee, \top, \mathbf{t}, \bot\}$ is functionally complete for $MSPL_\Delta$.

Here $\bot$ has been included, just because it is used to define $MSPL_\Delta$'s 'official' intuitionistic minimal negations. Since nothing particular is assumed about $\bot$ in $MSPL_\Delta$, the essential inventory is $\{/, \backslash, \wedge, \circ, \vee, \top, \mathbf{t}\}$. The absence of negations from this set is a remarkable fact, because apparently negation plays no role for definability. This is in sharp contrast to the functional completeness results for substructural subsystems of Nelson's propositional logic $\mathbf{N}^-$ in Chapter 7.

4.5 Digression: On the expressiveness of Categorial Grammar

In this section we shall address a central development in recent investigations of Categorial Grammar, viz. the introduction of new type forming operations (see e.g. [Moortgat 1988, 1990], [Morrill 1990]). If the parsing mechanism of Categorial Grammar is thought of as a propositional logic, the problem is to motivate and to characterize additional propositional connectives besides the two implications / and \ (and sometimes an additional juxtaposition connective). In what follows, a number of additional operations for the Lambek Calculus of syntactic types is motivated and positive sequential propositional logic '*PSPL*', i.e. positive propositional logic without structural rules of inference, is suggested as an extended syntactic calculus. By the results of the previous section, we obtain a functional completeness result for the connectives of the extended syntactic calculus *PSPL*.

4.5.1 Extended Lambek Calculus

The functor-argument structure of languages The standard bi-directional Ajdukiewicz Calculus of syntactic types [Ajdukiewicz 1935] makes use of (i) a finite number of basic type symbols like n (for names) and s (for sentences) and (ii) the directional implication signs / (right residuation) and \ (left residuation) which are used to build up functor types. In this propositional language of types simple natural language sentences can be given a categorial analysis by means of (directional versions of) Ajdukiewicz's cancellation rule; consider e.g.the following type assignments and natural deduction proof-trees:

(1) `Suddenly` `Mary` `discovered` `John`

$$\dfrac{(s/s) \qquad \dfrac{n \qquad \dfrac{((n\backslash s)/n) \quad n}{(n\backslash s)}}{s}}{s}$$

(2) `Suddenly` `Mary` `discovered` `John`

$$\dfrac{(s/s) \qquad \dfrac{\dfrac{n \quad (n\backslash(s/n))}{(s/n)} \quad n}{s}}{s}$$

An essential extension of Ajdukiewicz's Categorial Grammar has been developed by Lambek [1958]. The core of Lambek's generalization of Ajdukiewicz's syntactic calculus is to supplement the cancellation principles, which are directional versions of the modus ponens rule (or the principle of functional application), by the corresponding *conditionalization* (or functional abstraction) principles. In a sequent-style presentation, the resulting implicational logic turns out to be just intuitionistic (or positive) implicational logic without structural rules of inference. Thus, Ajdukiewicz's extremely simple parsing mechanism has been extended by Lambek into a full-fledged implicational logic, or, as many linguists prefer to say, standard Categorial Grammar has been developed into *flexible* Categorial Grammar. Categorial Grammar based on the Lambek Calculus is called flexible, because it gives rise to "derivational polymorphism", i.e. certain multiple type assignments can now be taken into account by the richer deductive means. The different syntactic types assigned to `discovered` in (1) and (2) e.g. are interderivable in the Lambek Calculus; for one direction see e.g. the following sequent-calculus proof:

$$\dfrac{\dfrac{\dfrac{\dfrac{n \to n \quad s \to s}{n(n \backslash s) \to s} \qquad n \to n}{n((n \backslash s)/n)n \to s}}{n((n \backslash s)/n) \to (s/n)}}{((n \backslash s)/n) \to (n \backslash (s/n))}$$

Ideally one would like to have the following picture: for any syntactic item the type assignment should be as parsimonious as possible; every additional type in which the item may occur should be derivable from an assigned type using the underlying syntactic calculus. What is important for our considerations here is that at this stage we have the following set of type-forming operations: $\{/, \backslash\}$.

Concatenation of syntactic types Lambek [1958] also introduces a conjunction operation denoting juxtaposition (or concatenation) of syntactic types in the object language. Let us use $\circ$ as the juxtaposition connective. Natural concatenative syntactic types are provided by texts. E.g. in the Lambek Calculus one can prove that the two sequences $n\ (n \backslash s)$ and $n\ ((n \backslash s)/n)\ n$ both derive s: $n\ (n \backslash s) \to s$, $n\ ((n \backslash s)/n)\ n \to s$. Now, one may put these sequences together to obtain a short text consisting of two sentences: $n\ (n \backslash s)\ n\ ((n \backslash s)/n)\ n \to (s \circ s)$. Thus, we arrive at the following set of operations: $\{/, \backslash, \circ\}$.

Non-derivational type ambiguity There exist type ambiguities which cannot be captured by the derivational apparatus. A well-known example is the type membership of the connective `and`. `And` does not only act as a propositional connective but it may also conjoin e.g. names as well as adverbs, as witnessed by the following two examples.

(3) `Peter and Mary discovered John`

$$\dfrac{\dfrac{n \qquad \dfrac{((n \backslash n)/n)\ n}{(n \backslash n)}}{n} \qquad \dfrac{((n \backslash s)/n)\ \ n}{(n \backslash s)}}{s}$$

(4) `Suddenly and unexpectedly John died`

$$\dfrac{\dfrac{(s/s) \qquad \dfrac{(((s/s) \backslash (s/s))/(s/s))\ \ (s/s)}{((s/s) \backslash (s/s))}}{(s/s)} \qquad \dfrac{n \qquad (n \backslash s)}{s}}{s}$$

The prevailing reaction in the literature to non-derivational type ambiguity is the introduction of polymorphic types, i.e. types involving type variables (cf. e.g. [van Benthem 1991]). `And` would thus be assigned the polymophic type $((x \backslash x)/x)$. One need not, however, resort to a higher-order framework in order to deal with the type ambiguitiy of expressions like `and`. Moreover, the descriptive adequacy of "variable polymorphism" may be questioned. `And` e.g. is not a conjunction of expressions in each syntactic type, as sentences like the following are clearly ungrammatical: `Peter and and and Mary discovered John`, `This is not and not my book`. In [Lambek 1961] one can find the operation $\cap$ of type intersection, which we shall denote by $\wedge$ to emphasize the logical perspective on Categorial Grammar. Instead of using type variables one could simply try to assign to and the type of a finite conjunction: **and** $\mapsto ((n\backslash n)/n)\wedge((s\backslash s)/s)\wedge\ldots$. At this stage we have already a set of four type forming operations, viz. two directional implications and two conjunctions (one in the sense of concatenation and one in the sense of type intersection): $\{/, \backslash, \circ, \wedge\}$.

Incomplete syntactic information Suppose for a moment that you are a linguist who has already succesfully carried out a partial categorial analysis of a certain maybe not yet syntactically investigated language. The language is a written language, and part of the dictionary you are compiling is the following type assignment: |||| $\mapsto ((n \backslash s)/n)\wedge(s/n)$; ©©© $\mapsto n$. Unfortunately, you don't know to which type the expression − − − − − belongs; you managed, however, to find out that the string − − − − − |||| ©©© is a sentence, i.e. we have the following situation:

(5)	− − − − −	||||	©©©	
	?	$((n \backslash s)/n) \wedge (s/n)$	n	$\rightarrow s.$

Now this incomplete information offers you some reasonable options for a type assignment to − − − − −, besides brute force assignments like − − − − − $\mapsto ((s/n)/(((n \backslash s)/n) \wedge (s/n)))$. Depending on whether |||| acts as an intransitive verb (s/n) or as a transitive verb $((n \backslash s)/n)$, you may assign the type (s/s) or the type n to − − − − −. Since both are possible, you finally decide to assign the disjunctive type $((s/s) \vee n)$. Assuming incomplete syntactic information thus naturally leads us to introducing disjunctive types. Our set of operations now looks like this: $\{/, \backslash, \circ, \wedge, \vee\}$.[3]

To a certain extent there is a difference in character between the connectives $/$, $\backslash$, and $\circ$ on the one hand and $\wedge$ and $\vee$ on the other hand. Whereas $/$, $\backslash$, and $\circ$ are directly related to *the process of parsing* ($/$ and $\backslash$ reflect the functorial structure to be found in languages; $\circ$ reflects the linear arrangement of linguistic items), $\wedge$ and $\vee$ are related to *the process of assigning types*. However, once types involving $\wedge$ and $\vee$ are assigned, applications of the rules governing $\wedge$ and $\vee$ are, of course, steps in the parsing process. Now that we have argued for an extended set of type-forming operations, a number of interesting questions arises:

1. Are there further reasonable operations for Categorial Grammar?
2. Are there sets of operations which allow one to define all possibilities?
3. Are there finite such sets?

Continuing our logical perspective, these questions may be paraphrased in more logical terms. What the first question essentially amounts to is the question whether there is a *semantics* and therefore a definitional framework telling us what is a possible operation. The second question can be translated as the problem of *functional completeness* for a given logic $\mathcal{L}$. In order to approach a functional completeness result for extended Categorial Grammar we thus have to specify a definitional framework and, of course,

[3] Kanazawa [1992] studies the effects on recognizing power of adding $\wedge$ and $\vee$ (his notation is '$\cap$' resp. '$\cup$') to the associative Lambek Calculus with product.

we have to syntactically characterize the informally motivated new connectives. It turns out that we can directly apply our earlier results.

Definition 4.13 Positive sequential propositional logic *PSPL* is the $\bot$-free part of *MSPL*.

Now, *PSPL* will serve as our extended syntactic calculus. The propositional constants **t** and $\top$ receive a straightforward linguistic interpretation: **t** denotes any sequence of type symbols, and $\bot$ denotes the empty sequence. In *PSPL* we assume denumerably many propositional variables (or basic type symbols), but this assumption is not essential. As an immediate consequence of our earlier results we obtain the following

Corollary 4.14 $\{/, \backslash, \wedge, \circ, \vee, \top, \mathbf{t}\}$ is functionally complete for $PSPL_\Delta$.

4.5.2 Limitations

The question of what is the philosophical significance of the previous theorem and of functional completeness results in general is worth a seperate treatment, and we will come back to it in Chapter 9. For the time being let us just note that the proof-theoretic framework we have used puts some heavy restrictions on definability. In fact there are interesting type-forming operations which cannot be defined in it. Consider e.g. the standard set-theoretic interpretation of /, \ and $\circ$:

$$\begin{aligned}(A/B) &= \{y \in V^+ \mid (\forall x \in B)\, yx \in A\},\\ (B \backslash A) &= \{y \in V^+ \mid (\forall x \in B)\, xy \in A\},\\ (A \circ B) &= \{z \in V^+ \mid (\exists x \in A)(\exists y \in B)\, xy = z\},\end{aligned}$$

where V^+ is the set of all non-empty, finite strings over a given vocabulary V. In order to handle certain syntactic discontinuity phenomena, Moortgat [1988] introduces the type-forming operations 'extraction' $\uparrow$ and 'infixation' $\downarrow$:

$$\begin{aligned}(A \uparrow B) &= \{xz \in V^+ \mid \forall y(y \in B \supset xyz \in A)\},\\ (A \downarrow B) &= \{y \in V^+ \mid \forall x \forall z(xz \in B \supset xyz \in A)\},\end{aligned}$$

and shows that $\uparrow$ and $\downarrow$ cannot be characterized by operational rules in an ordinary sequent calculus. They can be characterized, if in addition to sequent rules one uses what Moortgat calles "string equations" [Moortgat 1990]. Thus, if one wants to make use of type-forming operations like $\uparrow$ and $\downarrow$, this requires a proof-theoretic framework which is essentially richer than the above higher-lever Gentzen semantics.

Chapter 5

Formulas-as-types for substructural subsystems of *IPL*

In this chapter, Howard's [1969] 'formulas-as-types notion of construction' for intuitionistic implicational logic $IPL_{\supset}$, i.e. an encoding of proofs in $IPL_{\supset}$ by lambda terms, is extended to certain fragments of the substructural subsystems of *IPL* introduced in Chapter 3. This is achieved by taking up Buszkowski's [1987, 1988] distinction between two kinds of lambda-abstractors, singling out suitable fragments of typed terms (cf. also [van Benthem 1986], [Buszkowski 1987]), and appropriately modifying the notion of construction. The relationship between cut-elimination and normalization of terms is dealt with. Amongst other things, it is shown that in certain cases in which applications of the (*cut*)-rule can be eliminated from an implicational fragment of the logics considered, cut-elimination and normalization of terms wrt β-reduction are homomorphic images of each other. [1]

5.1 The typed lambda calculus $\lambda_{/,\backslash}$

We introduce a directional variant $\lambda_{/,\backslash}$ of the ordinary typed λ-calculus $\lambda_{\supset}$. The vocabulary of the term-language $T_{/,\backslash}$ consists of denumerably many term variables $v_1, v_2, \ldots$, every formula in $\{/, \backslash\}$, the lambda abstractors λ^r, λ^l, and brackets (,).

Definition 5.1 The set $\Lambda_{/,\backslash}$ of $T_{/,\backslash}$-terms is the smallest set Γ such that

(i) $V_{/,\backslash} =_{\text{def}} \{v_i^A \mid 0 < i \in \omega, A \text{ is a formula in } \{/, \backslash\}\} \subseteq \Gamma$;

(ii) if $M^A, N^{(B/A)} \in \Gamma$, then $(NM)^B \in \Gamma$;

(iii) if $M^A, N^{(A\backslash B)} \in \Gamma$, then $(MN)^B \in \Gamma$;

(iv) if $M^B \in \Gamma$, $x^A \in V_{/,\backslash}$, then $(\lambda^r xM)^{(B/A)}, (\lambda^l xM)^{(A\backslash B)} \in \Gamma$.

[1]The present chapter is based on [Wansing 1992]. [Gabbay & de Queiroz 1992] deals with exactly the same topic, but within a natural deduction framework. Gabbay and de Queiroz consider a range of formal systems which includes also classical 2-valued implicational logic. However, they do not distinguish between a right-searching and a left-searching implication. Moreover, they are not concerned with the relationship between certain operations on typed terms and operations on proofs like normalization wrt β-reduction and cut-elimination.

M^A is called a term of type A. Sometimes outermost parentheses and type-symbols of terms will be omitted. We shall use $x, y, z, w, x_1, x_2, \ldots$ etc. to denote elements from $V_{/,\backslash}$ and $M, N, G, H, M_1, M_2, \ldots$ etc. as to denote $T_{/,\backslash}$-terms. The set $FV(M)$ of free variables of M, the set $ST(M)$ of subterms of M, and $M[x^B := N^B]$, the result of substituting N for the occurrences of $x \in FV(M)$ in M, are inductively defined in the obvious way. A variable x in M is bound iff $x \notin FV(M)$. M is closed iff $FV(M) = \emptyset$. An occurrence of a free variable will be called a free variable-occurrence (fvo). $M \equiv N$ expresses that M, N are the same or are obtainable from each other by renaming bound variables. $\lambda^r x_1 \ldots x_n.M \equiv (\lambda^r x_1(\lambda^r x_2(\ldots(\lambda^r x_n M)\ldots)$, $\lambda^l x_1 \ldots x_n.M \equiv (\lambda^l x_1(\lambda^l x_2(\ldots(\lambda^l x_n M)\ldots)$.

Let $M[x^B]$ denote a $T_{/,\backslash}$-term in which x occurs as a free variable, and let $M[N^B]$ be the result of substituting N for a single occurrence of x in $M[x]$. It will always be clear from the context for which occurrence of x in $M[x]$ the term N is substituted to obtain $M[N]$. N is free for x in $M[x]$ iff no $y \in FV(N)$ becomes bound in $M[N]$.

Lemma 5.2 Instead of clauses (ii), (iii) in previous definition, one may equivalently use clauses

(ii)' if $M^C[x^B], N^A \in \Gamma$, $z^{(B/A)} \in V_{/,\backslash}$, then $M[zN] \in \Gamma$, provided zN is free for x in $M[x]$;

(iii)' if $M^C[x^B], N^A \in \Gamma$, $z^{(A\backslash B)} \in V_{/,\backslash}$, then $M[Nz] \in \Gamma$, provided Nz is free for x in $M[x]$;

(v) if $M^C[x^B], N^B \in \Gamma$, then $M[N] \in \Gamma$, provided N is free for x in $M[x]$.

PROOF (ii) implies (ii)': Suppose that $z^{(B/A)} \in V_{/,\backslash}, M^C[x^B], N^A \in \Lambda_{/,\backslash}$. By (ii), $zN \in \Lambda_{/,\backslash}$. By renaming fvos of type B in $M[x]$, $M[x][x := zN] \equiv M[zN] \in \Lambda_{/,\backslash}$. (iii) implies (iii)': analogously. (ii)', (v) imply (ii): Suppose that $N^{(B/A)}, M^A \in \Lambda_{/,\backslash}$, and $z^{(B/A)}, y^A, x^B \in V_{/,\backslash}$. By (ii)' and (v), $zy \in \Lambda_{/,\backslash}$. But then, by (v), $NM \in \Lambda_{/,\backslash}$. (iii)', (v) imply (iii): analogously. If (ii), (iii) are used, (v) follows by renaming fvos of the same type. □

The remaining part of this section and similar passages in later sections of this chapter are standard applications from (typed) lambda calculus.

Definition 5.3 The logical axiom-schemata and rules of $\lambda_{/,\backslash}$ are:

$M^A = M^A$;

if $M = N$, then $N = M$;

if $M = N, N = G$, then $M = G$;

if $M^{(A\backslash B)} = N^{(A\backslash B)}$, then $G^A M = GN$;

if $M^A = N^A$, then $MG^{(A\backslash B)} = NG$;

if $M^{(B/A)} = N^{(B/A)}$, then $MG^A = NG$;

if $M^A = N^A$, then $G^{(B/A)}M = GN$;

if $M = N$, then $\lambda^r xM = \lambda^r xN$;

if $M = N$, then $\lambda^l xM = \lambda^l xN$.

The axiom-schemata of $\lambda_{/,\backslash}$'s theory of typed β-equality are:

(β^r) $(\lambda^r x^A.M)N^A = M\,[x := N]$;

(β^l) $N^A(\lambda^l x^A.M) = M\,[x := N]$.

Definition 5.4 The binary relations $\rightarrowtail_\beta$ (one-step β-reduction), $\twoheadrightarrow_\beta$ (β-reduction), and $=_\beta$ (β-convertability) are defined as follows:

(1) $(\lambda^r x^A.M)N^A \rightarrowtail_\beta M\,[x := N]$; $N^A(\lambda^l x^A.M) \rightarrowtail_\beta M\,[x := N]$;

if $M^A \rightarrowtail_\beta N^A$, then $MG^{(A\backslash B)} \rightarrowtail_\beta NG^{(A\backslash B)}$, $G^{(B/A)}M \rightarrowtail_\beta G^{(B/A)}N$;

if $M^{(A\backslash B)} \rightarrowtail_\beta N^{(A\backslash B)}$, then $G^A M \rightarrowtail_\beta GN$;

if $M^{(B/A)} \rightarrowtail_\beta N^{(B/A)}$, then $MG^A \rightarrowtail_\beta NG$;

if $M \rightarrowtail_\beta N$, then $\lambda^r xM \rightarrowtail_\beta \lambda^r xN$, $\lambda^l xM \rightarrowtail_\beta \lambda^l xN$;

(2) $\twoheadrightarrow_\beta$ is the reflexive and transitive closure of $\rightarrowtail_\beta$;

(3) $=_\beta$ is the equivalence relation generated by $\twoheadrightarrow_\beta$.

Definition 5.5 $T_{/,\backslash}$-terms $(\lambda^r x^A.M)N^A$, $N^A(\lambda^l x^A.M)$ are called β-redexes. $M\,[x := N]$ is called the contractum of both $(\lambda^r x^A.M)N^A$ and $N^A(\lambda^l x^A.M)$. M is a β-normal form (β-nf) iff it has no β-redexes as a subterm. M has a β-nf iff there exists a N such that $M =_\beta N$ and N is a β-nf.

Theorem 5.6 (Church-Rosser Theorem) If $M \twoheadrightarrow_\beta N_1, M \twoheadrightarrow_\beta N_2$, then there exists an N_3 such that $N_1 \twoheadrightarrow_\beta N_3, N_2^A \twoheadrightarrow_\beta N_3$.

PROOF Like the proof of the Church-Rosser Theorem 3.2.8. (i) in [Barendregt 1984]. □

Using the Church-Rosser Theorem it can be shown that every M has at most one β-nf. Moreover, as in the case of the undirectional typed lambda calculus $\lambda_\supset$, one can prove a strong normalization theorem for $\lambda_{/,\backslash}$, implying that every M in fact has a β-nf.

Definition 5.7 M is called strongly normalizable (sn) wrt $\twoheadrightarrow_\beta$ iff every β-reduction starting at M is finite.

Theorem 5.8 Every M is sn wrt $\twoheadrightarrow_\beta$.

PROOF See Appendix 5.8. □

Thus, every M has exactly one β-nf. Let $NORM_\beta$ denote iterated contraction of the leftmost β-redex. Since every β-reduction starting at any M is finite, $NORM_\beta$ constitutes a terminating normalization algorithm wrt $\twoheadrightarrow_\beta$.

5.2 Encoding proofs in $ISPL_{/,\backslash}$

Let $ISPL_{/,\backslash}$ denote the /, \-fragment of $ISPL$, and let $PROOF_{ISPL_{/,\backslash}}$ denote the set of proofs in $ISPL_{/,\backslash}$.[2] Next, we shall define a fragment of $\Lambda_{/,\backslash}$ such that every $\Pi \in PROOF_{ISPL_{/,\backslash}}$ can be encoded by some term from this fragment, and vice versa.

Definition 5.9 Let $\Lambda_{ISPL_{/,\backslash}}$ be the biggest $\Gamma \subseteq \Lambda_{/,\backslash}$ such that

(i) for every $M \in \Gamma$ and every prefix of the form $\lambda^r x$ resp. $\lambda^l x$ in M, $\lambda^r x$ resp. $\lambda^l x$ binds exactly one fvo;

(ii) for every $\lambda^r x.M \in \Gamma$, an occurrence of x is the rightmost fvo in M;

(iii) for every $\lambda^l x.M \in \Gamma$, an occurrence of x is the leftmost fvo in M.

Note that $\Lambda_{ISPL_{/,\backslash}}$ is closed under substitutions of a term N for a fvo x in a term M, since N must be free for x in $M[x]$.

Definition 5.10 A term $M^B \in \Lambda_{ISPL_{/,\backslash}}$ is a construction of a sequent $A_1 \dots A_n \to B$ iff M has exactly fvos $x_1^{A_1}, \dots, x_n^{A_n}$, in this order from left to right.

Theorem 5.11 Given a proof in $PROOF_{ISPL_{/,\backslash}}$ of a sequent $s = A_1 \dots A_n \to B$, one can find a construction $M^B \in \Lambda_{ISPL_{/,\backslash}}$ of s, and conversely.

Proof We shall inductively define encoding functions $f : PROOF_{ISPL_{/,\backslash}} \longrightarrow \Lambda_{ISPL_{/,\backslash}}$, $g : \Lambda_{ISPL_{/,\backslash}} \longrightarrow PROOF_{ISPL_{/,\backslash}}$, such that it can readily be seen that $f(\Pi)$ is a construction of Π, and $g(M)$ proves a sequent of which M is a construction. Let $(N)^\natural$ denote the result of renaming (from left to right) the fvos in N by occurrences of *distinct* variables of the respective types and with smallest possible indices such that the renamed variable occurrences are free in N. Moreover, for $M[x^B]$, $(N^B)^\natural$ is required to be free for x in $M[(N)^\natural]$. We shall use $\Pi, \Pi_1, \Pi_2, \dots$, to denote proofs. The function f is inductively defined as follows:

- $\Pi = A \to A$: $f(\Pi) \equiv v_1^A$;
- $\Pi = \frac{\frac{\Pi_1}{XA \to B}}{X \to (B/A)}$: $f(\Pi) \equiv \lambda^r v_i^A . f(\frac{\Pi_1}{XA \to B})$, where an occurrence of v_i is the rightmost fvo of type A in $f(\frac{\Pi_1}{XA \to B})$;
- $\Pi = \frac{\frac{\Pi_1}{Y \to A} \quad \frac{\Pi_2}{XBZ \to C}}{X(B/A)YZ \to C}$: $f(\Pi) \equiv (f(\frac{\Pi_2}{XBZ \to C})[(v_j^{(B/A)} f(\frac{\Pi_1}{Y \to A}))^\natural])^\natural$, where v_j is the first variable of type (B/A) not occurring in $f(\frac{\Pi_1}{Y \to A})$;
- for $\Pi = \frac{\frac{\Pi_1}{AX \to B}}{X \to (A \backslash B)}$ resp. $\Pi = \frac{\frac{\Pi_1}{Y \to A} \quad \frac{\Pi_2}{XBZ \to C}}{XY(A \backslash B)Z \to C}$, $f(\Pi)$ is defined in analogy to the previous two cases;
- $\Pi = \frac{\frac{\Pi_1}{Y \to A} \quad \frac{\Pi_2}{XAZ \to B}}{XYZ \to B}$: $f(\Pi) = (f(\frac{\Pi_2}{XAZ \to B})[(f(\frac{\Pi_1}{Y \to A}))^\natural])^\natural$.

[2]Obviously, $ISPL_{/,\backslash} = MSPL_{/,\backslash}$.

The function g is inductively defined as follows:

- $M \equiv x^B$: $g(M) = B \to B$;
- $M \equiv \lambda^r x^A.N^C$: $g(M) = \frac{g(N)}{X \to (C/A)}$,
where XA is the sequence of types of the fvos in N;
- $M \equiv N^B\,[z^{(C/A)}G^A]$: $g(M) = \frac{g(G)\quad g(N\,[v_i^C])}{X(C/A)YZ \to B}$,
where Y is the sequence of types of the fvos in G, XCZ is the sequence of types of the fvos in $N\,[v_i]$, and v_i is the first variable of type C not occurring in N;
- for $M \equiv \lambda^l x^A.N^C$ resp. $M \equiv N^B\,[G^A z^{(A\backslash C)}]$, $g(M)$ is defined in analogy to the previous two cases;
- $M \equiv N^B\,[H^C]$ and H^C is neither a variable nor of the form $G^A z^{(A\backslash C)}$ or $z^{(C/A)}G^A$: $g(M) = \frac{g(H)\quad g(N\,[v_i^C])}{XYZ \to B}$, where Y is the sequence of types of the fvos in H, XCZ is the sequence of types of the fvos in $N\,[v_i]$, and v_i is the first variable of type C not occurring in N. □

For a different proof of this theorem wrt to a different sequent calculus for $ISPL_{/,\backslash}$, see the proof of Theorem 1.5 in [Buszkowski 1987]. From the definition of $\Lambda_{ISPL_{/,\backslash}}$ it becomes clear that for the encoding one may instead of λ^l and λ^r just use the ordinary lambda-abstractor λ together with suitable constraints on variable-binding. What is crucial, however, is the use of directional types.

5.3 Cut-elimination in $PROOF_{ISPL_{/,\backslash}}$ and β-reduction in $\Lambda_{ISPL_{/,\backslash}}$ as homomorphic images of each other

Let $ELIM_c$ denote the terminating cut-elimination algorithm for $ISPL_{/,\backslash}$ obtained by the proof of Theorem 3.4. We shall specify in which sense cut-elimination in $PROOF_{ISPL_{/,\backslash}}$ 'corresponds' to normalization wrt $\twoheadrightarrow_\beta$ in $\Lambda_{ISPL_{/,\backslash}}$. (For an analogous result wrt to cut-elimination on proofs in a symmetric sequent calculus for IPL and normalization of natural deduction proofs in IPL see [Pottinger 1977].)

Theorem 5.12 Let $\mathcal{A} =< PROOF_{ISPL_{/,\backslash}}, ELIM_c >$, $\mathcal{B} =< \Lambda_{ISPL_{/,\backslash}}, NORM_\beta >$.

(i) The function f defined in the proof of the previous theorem is a homomorphism from $\mathcal{A}$ to $\mathcal{B}$.

(ii) The function g defined in the proof of the previous theorem is a homomorphism from $\mathcal{B}$ to $\mathcal{A}$.

Proof First of all note that $\Lambda_{ISPL_{/,\backslash}}$ is closed under $NORM_\beta$, i.e. $NORM_\beta$ is in fact a function on $\Lambda_{ISPL_{/,\backslash}}$. (i): It has to be shown that $f(ELIM_c(\Pi)) \equiv NORM_\beta(f(\Pi))$. The cases in which the last step in the generation of Π is not an application of (cut) are straightforward. If $\Pi = \frac{\frac{\Pi_1}{Y \to A}\quad \frac{\Pi_2}{XAZ \to B}}{XYZ \to B}$, then in order to determine $f(ELIM_c(\Pi))$, one has to consider every conversion step involving / or \ in the proof of Lemma 3.3. The essential elimination steps in which one application of (*cut*) is replaced by two applications of (*cut*) of a smaller degree amount to one-step β-reductions between the respective f-images, whereas in each remaining step the f-images are the same:

$$\left[\dfrac{\dfrac{\dfrac{\Pi_1}{XA\to B}}{X\to(B/A)} \qquad \dfrac{\dfrac{\Pi_2}{Y_1\to A}\quad\dfrac{\Pi_3}{X_1BZ\to C}}{X_1(B/A)Y_1Z\to C}}{X_1XY_1Z\to C}\right] \text{ is converted into } \left[\dfrac{\dfrac{\dfrac{\Pi_2}{Y_1\to A}\quad\dfrac{\Pi_1}{XA\to B}}{XY_1\to B}\qquad\dfrac{\Pi_3}{X_1BZ\to C}}{X_1XY_1Z\to C}\right]$$

$$\downarrow f \qquad\qquad\qquad\qquad \downarrow f$$

$$\dfrac{\dfrac{f(\frac{\Pi_1}{XA\to B})}{\lambda^r v_i^A . f(\frac{\Pi_1}{XA\to B})} \quad \dfrac{f(\frac{\Pi_2}{Y_1\to A})\quad f(\frac{\Pi_3}{X_1BZ\to C})}{(f(\frac{\Pi_3}{X_1BZ\to C})[((v_j f(\frac{\Pi_2}{Y_1\to A}))^B)^\natural])^\natural}}{((f(\frac{\Pi_3}{X_1BZ\to C})[(v_j f(\frac{\Pi_2}{Y_1\to A}))^\natural])^\natural[(\lambda^r v_i^A.N)^\natural])^\natural} \twoheadrightarrow_\beta \dfrac{\dfrac{f(\frac{\Pi_1}{Y_2\to A})\quad f(\frac{\Pi_1}{XA\to B})}{(f(\frac{\Pi_1}{XA\to B})[(f(\frac{\Pi_2}{Y_1\to A}))^\natural])^\natural}\quad f(\frac{\Pi_3}{X_1BZ\to C})}{(f(\frac{\Pi_3}{X_1BZ\to C})[(N[(f(\frac{\Pi_2}{Y_1\to A}))^\natural])^\natural])^\natural},$$

where $N \equiv f(\frac{\Pi_1}{XA\to B})$;

$$\left[\dfrac{\dfrac{\dfrac{\Pi_1}{Y_1\to C}\quad\dfrac{\Pi_2}{X_1DZ_1\to A}}{X_1(D/C)Y_1Z_1\to A}\qquad\dfrac{\Pi_3}{XAZ\to B}}{XX_1(D/C)Y_1Z_1Z\to B}\right] \text{ is converted into } \left[\dfrac{\dfrac{\Pi_1}{Y_1\to C}\qquad\dfrac{\dfrac{\Pi_2}{X_1DZ_1\to A}\quad\dfrac{\Pi_3}{XAZ\to B}}{XX_1DZ_1Z\to B}}{XX_1(D/C)Y_1Z_1Z\to B}\right]$$

$$\downarrow f \qquad\qquad\qquad\qquad \downarrow f$$

$$\dfrac{\dfrac{f(\frac{\Pi_1}{Y_1\to C})\quad N}{(N[(v_j^{(D/C)} f(\frac{\Pi_1}{Y_1\to C}))^\natural])^\natural}\quad f(\frac{\Pi_3}{XAZ\to B})}{(f(\frac{\Pi_3}{XAZ\to B})[(N[(v_j f(\frac{\Pi_1}{Y_1\to C}))^\natural])^\natural])^\natural} \equiv \dfrac{f(\frac{\Pi_1}{Y_1\to C})\qquad\dfrac{f(\frac{\Pi_2}{X_1DZ_1\to A})\quad N}{(f(\frac{\Pi_3}{XAZ\to B})[(N)^\natural])^\natural}}{((f(\frac{\Pi_3}{XAZ\to B})[(N)^\natural])^\natural[(v_j f(\frac{\Pi_1}{Y_1\to C}))^\natural])^\natural},$$

where $N = f(\frac{\Pi_2}{X_1DZ_1\to A})$;

$$\left[\dfrac{\dfrac{\Pi_1}{Y\to A}\qquad\dfrac{\dfrac{\Pi_2}{XAZC\to B}}{XAZ\to(B/C)}}{XYZ\to(B/C)}\right] \text{ is converted into } \left[\dfrac{\dfrac{\dfrac{\Pi_1}{Y\to A}\quad\dfrac{\Pi_2}{XAZC\to B}}{XYZC\to B}}{XYZ\to(B/C)}\right]$$

$$\downarrow f \qquad\qquad\qquad\qquad \downarrow f$$

$$\dfrac{f(\frac{\Pi_1}{Y\to A})\quad\dfrac{f(\frac{\Pi_2}{XAZC\to B})}{\lambda^r v_i^C.f(\frac{\Pi_2}{XAZC\to B})}}{(\lambda^r v_i^C.f(\frac{\Pi_2}{XAZC\to B})[(f(\frac{\Pi_1}{Y\to A}))^\natural])^\natural} \equiv \dfrac{\dfrac{f(\frac{\Pi_1}{Y\to A})\quad f(\frac{\Pi_2}{XAZC\to B})}{(f(\frac{\Pi_2}{XAZC\to B})[(f(\frac{\Pi_1}{Y\to A}))^\natural])^\natural}}{\lambda^r v_i^C.(f(\frac{\Pi_2}{XAZC\to B})[(f(\frac{\Pi_1}{Y\to A}))^\natural])^\natural};$$

$$\left[\;\dfrac{\frac{\Pi_1}{Y\to A}\qquad \dfrac{\frac{\Pi_2}{XAZ_1CZ_2\to B}\qquad \frac{\Pi_3}{Y_1\to D}}{XAZ_1(C/D)Y_1Z_2\to B}}{XYZ_1(C/D)Y_1Z_2\to B}\;\right] \text{ is converted into } \left[\;\dfrac{\dfrac{\frac{\Pi_1}{Y\to A}\qquad \frac{\Pi_2}{XAZ_1CZ_2\to B}}{XYZ_1CZ_2\to B}\qquad \frac{\Pi_3}{Y_1\to D}}{XYZ_1(C/D)Y_1Z_2\to B}\;\right]$$

$$\downarrow f \qquad\qquad\qquad\qquad \downarrow f$$

$$\dfrac{f(\frac{\Pi_1}{Y\to A})\quad \dfrac{f(\frac{\Pi_2}{XAZ_1CZ_2\to D})\quad f(\frac{\Pi_3}{Y_1\to D})}{(f(\frac{\Pi_2}{XAZ_1CZ_2\to B})[(v_j^{(C/D)}f(\frac{\Pi_3}{Y_1\to D}))^\natural])^\natural}}{((f(\frac{\Pi_2}{XAZ_1CZ_2\to B})[(v_jf(\frac{\Pi_3}{Y_1\to D}))^\natural])^\natural[(N)^\natural])^\natural} \equiv \dfrac{\dfrac{f(\frac{\Pi_1}{Y\to A})\quad f(\frac{\Pi_2}{XAZ_1CZ_2\to B})}{(f(\frac{\Pi_2}{XAZ_1CZ_2\to B})[(f(\frac{\Pi_1}{Y\to A}))^\natural])^\natural}\quad f(\frac{\Pi_3}{Y_1\to D})}{((f(\frac{\Pi_2}{XAZ_1CZ_2\to B})[(N)^\natural])^\natural[(v_jf(\frac{\Pi_3}{Y_1\to D}))^\natural])^\natural},$$

where $N \equiv f(\frac{\Pi_1}{Y\to A})$.

By the ind. hyp., $f(ELIM_c(\frac{\Pi_1}{Y\to A})) \equiv NORM_\beta(f(\frac{\Pi_1}{Y\to A}))$ and $f(ELIM_c(\frac{\Pi_2}{XAZ\to B})) \equiv NORM_\beta(f\frac{\Pi_2}{XAZ\to B}))$. Thus, $f(ELIM_c(\Pi)) \equiv NORM_\beta(f(\Pi))$. (ii): It has to be shown that $g(NORM_\beta(M)) \equiv ELIM_c(g(M))$. The cases in which the last step in the generation of M is not an application of clause (v) in Lemma 5.2. are straightforward. If $M \equiv N^B[H^C]$, then one has to distinguish among four cases and a number of subcases which together are exhaustive. Case 1: $H \equiv \lambda^r x^A.G^C$. (*a*) M is obtained from $x^{(C/A)}$. Then $g(NORM_\beta(M)) = g(NORM_\beta(H)) = ELIM_c(g(M))$ by (ii). (*b*) M is obtained from $z^{(C/A)}N_1^A$. Then either (i) $g(NORM_\beta(M)) = g(NORM_\beta(zN_1[H])) = ELIM_c(g(M))$ by (iii), or (ii) $g(NORM_\beta(M)) = g(NORM_\beta(HN_1)) = g(NORM_\beta(G[x^A := N_1])) = g(NORM_\beta(G[N_1])) = ELIM_c(g(M))$ by the induction hypothesis. (*c*) M is obtained from $\lambda^r y^D.G^{D_1}$. Then $g(NORM_\beta(M)) = ELIM_c(g(M))$ by (ii). (*d*) M is obtained from $\lambda^l y^D.G^{D_1}$: analogous to the previous subcase. Case 2: $H \equiv \lambda^l x^A.G^C$: analogous to Case 1. Case 3: $H \equiv (\lambda^r x^A.G^C)G_1^A$. Then $g(NORM_\beta(M)) = g(NORM_\beta(N[G[x^A := G_1^A]]))$ $= g(NORM_\beta(N[G[G_1^A]])) = ELIM_c(g(M))$ by the induction hypothesis. Case 4: $H \equiv G_1^A(\lambda^r x^A.G^C)$: analogous to Case 3. □

REMARK (i) There are no encoding functions $f' : PROOF_{ISPL_{/,\backslash}} \longrightarrow \Lambda_{ISPL_{/,\backslash}}$, $g' : \Lambda_{ISPL_{/,\backslash}} \longrightarrow PROOF_{ISPL_{/,\backslash}}$ which are monomorphisms from $\mathcal{A}$ to $\mathcal{B}$ resp. from $\mathcal{B}$ to $\mathcal{A}$: every variable of type A is in β-nf and encodes one and the same proof, viz. $A \to A$. Conversely, $A \to A$ and $\frac{A\to A \quad A\to A}{A\to A}$ e.g. are encoded by one and the same term in β-nf, viz. v_1^A.
(ii) Let $ELIM_c(PROOF_{ISPL_{/,\backslash}}) = \{ELIM_c(\Pi) \mid \Pi \in PROOF_{ISPL_{/,\backslash}}\}$, $NORM_\beta(\Lambda_{ISPL_{/,\backslash}}) = \{NORM_\beta(M) \mid M \in \Lambda_{ISPL_{/,\backslash}}\}$. By induction on $\Pi \in ELIM_c(PROOF_{ISPL_{/,\backslash}})$, it can easily be verified that $g(f(\Pi)) = \Pi$. Therefore $f \restriction ELIM_c(PROOF_{ISPL_{/,\backslash}})$ is a $1-1$-function from $ELIM_c(PROOF_{ISPL_{/,\backslash}})$ in $NORM_\beta(\Lambda_{ISPL_{/,\backslash}})$.

The previous theorem can be visualized as follows:

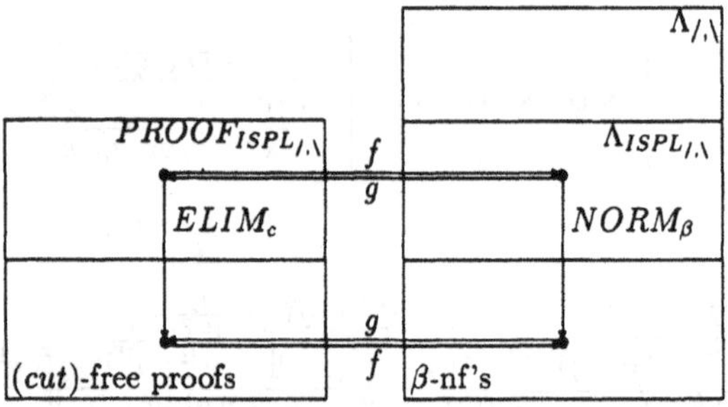

Figure 5.1: The relation between cut-elimination and β-reduction.

5.4 $\beta\eta$-reduction in $\Lambda_{ISPL_{/,\backslash}}$ as a monomorphic image of normalizing (cut)-free proofs

Definition 5.13 The axiom-schemata of $\lambda_{/,\backslash}$'s theory of typed η-equality are:

(η^r) $\lambda^r x^A.(M^{(B/A)}x^A) = M$, if $x \notin FV(M)$;

(η^l) $\lambda^l x^A.(x^A M^{(A\backslash B)}) = M$, if $x \notin FV(M)$.

The binary relations $\rightarrowtail_\eta$ (one-step η-reduction), $\twoheadrightarrow_\eta$ (η-reduction), and $=_\eta$ (η-convertability) are defined in the same way as one-step β-reduction, β-reduction, and β-convertability.

Definition 5.14 $T_{/,\backslash}$-terms $\lambda^r x^A.(M^{(B/A)}x^A)$, $\lambda^l x^A.(x^A M^{(A\backslash B)})$ are called η-redexes, if $x \notin FV(M)$. M is called the contractum of $\lambda^r x^A.(M^{(B/A)}x^A)$ resp. $\lambda^l x^A.(x^A M^{(A\backslash B)})$. M is an η-normal form (η-nf) iff it has no η-redex as a subterm. M has an η-nf iff there is an N such that $M =_\eta N$ and N is an η-nf.

Definition 5.15 If $M \rightarrowtail_\beta N$ or $M \rightarrowtail_\eta N$, then $M \rightarrowtail_{\beta\eta} N$ (M one-step $\beta\eta$-reduces to N).

The binary relations $\twoheadrightarrow_{\beta\eta}$ ($\beta\eta$-reduction) and $=_{\beta\eta}$ ($\beta\eta$-convertability) are defined in the same way as β-reduction and β-convertability.

Definition 5.16 M is a $\beta\eta$-normal form ($\beta\eta$-nf) iff it has neither a β-redex nor an η-redex as a subterm. M has a $\beta\eta$-nf iff there is an N such that $M =_{\beta\eta} N$ and N is a $\beta\eta$-nf.

Theorem 5.17 (Church-Rosser Theorem) If $M \twoheadrightarrow_{\beta\eta} N_1, M \twoheadrightarrow_{\beta\eta} N_2$, then there exists an N_3 such that $N_1 \twoheadrightarrow_{\beta\eta} N_3, N_2 \twoheadrightarrow_{\beta\eta} N_3$.

PROOF Like the proof of the Church-Rosser Theorem 3.3.9. (i) in [Barendregt 1984]. □

Theorem 5.18 M has a $\beta\eta$-nf iff it has a β-nf.

PROOF See the proof of corollary 15.1.5. in [Barendregt 1984]. □

Definition 5.19 M is called strongly normalizable (sn) wrt $\twoheadrightarrow_{\beta\eta}$ iff all $\beta\eta$-reductions starting at M are finite.

Theorem 5.20 Every M is sn wrt $\twoheadrightarrow_{\beta\eta}$.

PROOF See Appendix 5.8. □

Theorem 5.21 If $M \twoheadrightarrow_{\beta\eta} N$, then there is a G such that $M \twoheadrightarrow_{\beta} G \twoheadrightarrow_{\eta} N$.

PROOF See the proof of corollary 15.1.6. in [Barendregt 1984]. □

Let $NORM_\eta$ denote iterated contraction of the leftmost η-redex, and let $NORM$ denote $NORM_\eta NORM_\beta$, i.e. the composition of both algorithms. Since every $\beta\eta$-reduction starting at any M is finite, by the previous theorem, $NORM$ constitutes a terminating normalization algorithm wrt $\twoheadrightarrow_{\beta\eta}$. Now, call proofs of the following forms redundant:

$$\frac{\dfrac{A \to A \quad B \to B}{(B/A)\,A \to B}}{(B/A) \to (B/A)} \qquad \frac{\dfrac{A \to A \quad B \to B}{A\,(A \setminus B) \to B}}{(A \setminus B) \to (A \setminus B)}.$$

Lemma 5.22 Every proof of $X \to A$ in $PROOF_{ISPL_{/,\setminus}}$ can be converted into a proof of $X \to A$ in $PROOF_{ISPL_{/,\setminus}}$ without any redundant part.

PROOF By iterated application of the obvious conversions starting from the top. See e.g. the following conversion (together with the generation of the respective f-images):

$$\left[\frac{\dfrac{A \to A \quad B \to B}{(B/A)\,A \to B}}{(B/A) \to (B/A)}\right] \text{ is converted into } \quad [(B/A) \to (B/A)]$$

$$\downarrow f \qquad\qquad\qquad\qquad \downarrow f$$

$$\frac{\dfrac{v_1^A \quad v_1^B}{v_1^{(B/A)}\, v_1^A}}{\lambda^r v_1^A.(v_1^{(B/A)} v_1^A)} \quad \rightarrowtail_\eta \qquad\qquad v_1^{(B/A)}$$ □

Let $ELIM_r$ denote the (terminating) algorithm for eliminating redundant parts of proofs in $PROOF_{ISPL_{/,\setminus}}$, and let $ELIM$ denote $ELIM_r ELIM_c$, i.e. the composition of both algorithms.

Theorem 5.23 Let $\mathcal{A}_1 =< PROOF_{ISPL_{/,\setminus}}, ELIM >$; $\mathcal{B}_1 =< \Lambda_{ISPL_{/,\setminus}}, NORM >$.

(i) The function f defined in the proof of Theorem 5.11 is a monomorphism (i.e. a homomorphism which is $1-1$) from $\mathcal{A}_1$ to $\mathcal{B}_1$.

(ii) The function g defined in the proof of Theorem 5.11 is a homomorphism from $\mathcal{B}_1$ to $\mathcal{A}_1$.

PROOF Since $NORM_\beta(\Lambda_{ISPL_{/,\backslash}})$ is closed under $NORM_\eta$ and $ELIM_c(PROOF_{ISPL_{/,\backslash}})$ is closed under $ELIM_r$, $NORM$ is in fact a function on $\Lambda_{ISPL_{/,\backslash}}$ and $ELIM$ is a function on $PROOF_{ISPL_{/,\backslash}}$. (i): By Theorem 5.12 and the above remark, in order to prove that f is a $1-1$ homomorphism from $\mathcal{A}_1$ to $\mathcal{B}_1$ it is sufficient to show that $f \restriction ELIM_c(PROOF_{ISPL_{/,\backslash}})$ is a homomorphism from $< ELIM_c(PROOF_{ISPL_{/,\backslash}}), ELIM_r >$ in $< NORM_\beta(\Lambda_{ISPL_{/,\backslash}}), NORM_\eta >$. It has to be proved that $f \restriction ELIM_c(PROOF_{ISPL_{/,\backslash}})(ELIM_r(\Pi)) \equiv NORM_\eta(f \restriction ELIM_c(PROOF_{ISPL_{/,\backslash}})(\Pi))$. The proof by induction on the generation of $\Pi \in ELIM_c(PROOF_{ISPL_{/,\backslash}})$ is straightforward. (ii) It suffices to show by induction on the generation of $M \in NORM_\beta(\Lambda_{ISPL_{/,\backslash}})$ that $g \restriction NORM_\beta(\Lambda_{ISPL_{/,\backslash}})(NORM_\eta(M)) = ELIM_r(g \restriction NORM_\beta(\Lambda_{ISPL_{/,\backslash}})(M))$. □

5.5 Encoding proofs in $ISPL_{/,\backslash,\circ}$

We shall extend our considerations to the $/, \backslash, \circ$-fragment $ISPL_{/,\backslash,\circ}$ of $ISPL$. For this purpose we first define a set of typed terms $\Lambda_{/,\backslash,\circ}$. The vocabulary of the term-language $T_{/,\backslash,\circ}$ is the same as the vocabulary of $T_{/,\backslash}$ except that it contains every formula in $\{/, \backslash, \circ\}$ and the following new term-forming operators: $< -, - >$ (pairing) and $(-)_0$, $(-)_1$ (left and right projection).

Let $M\,[x^A, y^B]$ denote a $T_{/,\backslash,\circ}$-term with fvos x, y which are adjacent in the term's sequence of fvos, and let $M\,[G^A, H^B]$ be the result of (i) substituting G for the indicated occurrence of x in M and (ii) substituting H for the indicated occurrence of y in M. G, H are free for x, y in $M\,[x, y]$ iff no $z \in FV(G) \cup FV(H)$ becomes bound in $M\,[G, H]$.

Definition 5.24 The set $\Lambda_{/,\backslash,\circ}$ of $T_{/,\backslash,\circ}$-terms is the smallest set Γ such that

(i) $V_{/,\backslash,\circ} = \{v_i^A \mid 0 < i \in \omega,$ A is a formula in $\{/, \backslash, \circ\}\ \} \subseteq \Gamma$;

(ii) clauses (ii)', (iii)', (v) of Lemma 5.2 hold;

(iii) if M^A, $N^B \in \Gamma$, then $< M, N >^{(A \circ B)} \in \Gamma$;

(iv) if $M\,[x^A, y^B] \in \Gamma$, $z^{(A \circ B)} \in V_{/,\backslash,\circ}$, then $M\,[(z)_0, (z)_1] \in \Gamma$.

Note that we allow for projections not only of terms which are pairs (a fact that will be used in the proof of Theorem 5.32 below). If not otherwise stated, in the sequel we shall assume abbreviations, conventions, and definitions analogous to those introduced in connection with $\lambda_{/,\backslash}$. In particular, $FV(< M, N >) = FV(M) \cup FV(N)$; $FV((M)_i) = FV(M)$, $ST(< M, N >) = ST(M) \cup ST(N) \cup \{< M, N >\}$; $ST((M)_i) = ST(M) \cup \{(M)_i)\}$, $i = 0, 1$.

Definition 5.25 The binary relations $\succ$, $\succeq$, $\approx$ on $\Lambda_{/,\backslash,\circ}$ are defined as follows:

(1) $(\lambda^r x^A.M)N^A \succ M\,[x := N]$, $N^A(\lambda^l x^A.M) \succ M\,[x := N]$;

$(< M, N >)_0 \succ M$, $(< M, N >)_1 \succ N$;

if $M^A \succ N^A$, then $MG^{(A\backslash B)} \succ NG^{(A\backslash B)}$, $G^{(B/A)}M \succ G^{(B/A)}N$;

if $M^{(A\backslash B)} \succ N^{(A\backslash B)}$, then $G^A M \succ GN$;

if $M^{(B/A)} \succ N^{(B/A)}$, then $MG^A \succ NG$;

if $M \succ N$, then $\lambda^r x M \succ \lambda^r x N$, $\lambda^l x M \succ \lambda^l x N$;

if $M \succ N$, then $< M, G > \succ < N, G >$, $< G, M > \succ < G, N >$;

(2) $\succeq$ is the reflexive and transitive closure of $\succ$;

(3) $\approx$ is the equivalence relation generated by $\succeq$.

Definition 5.26 $T_{/,\backslash,\circ}$-terms $(\lambda^r x^A.M)N^A$, $N^A(\lambda^l x^A.M)$, $(< M, N >)_0$, $(< M, N >)_1$ are called redexes. $M[x := N]$ is called the contractum of both $(\lambda^r x^A.M)N^A$ and $N^A(\lambda^l x^A.M)$; M is called the contractum of $(< M, N >)_0$ and N is called the contractum of $(< M, N >)_1$. M is a normal form (nf) iff it has no redexes as a subterm. M has a nf iff there exists a N such that $M \approx N$ and N is a nf.

Theorem 5.27 (strong normalization) Every M is sn wrt $\succeq$.

PROOF Cf. the definition of computability and the proof of Theorem 2. in [de Vrijer 1987]. □

Theorem 5.28 (Church-Rosser Theorem) If $M \succeq N_1$, $M \succeq N_2$, then there exists an N_3 such that $N_1 \succeq N_3$, $N_2 \succeq N_3$.

PROOF In view of the previous theorem it is enough to establish the Church-Rosser property for sn terms. Cf. e.g. the proof of proposition 12.1. in [Lambek & Scott 1986]. □

Let $NORM_{\succeq}$ denote iterated contraction of the leftmost redex. $NORM_{\succeq}$ is a terminating normalization algorithm wrt $\succeq$.

Definition 5.29 ([Roorda 1991]) Let $N \in ST(M)$. N counts for one in M if either there is only one occurrence of N in M or there is a $G \in ST(M)$ such that N counts for one in G and $(G)_0$, $(G)_1$ count for one in M. If N is a variable, then N must be free in M in order to count for one.

Occurrences which together count for one will be identified and treated as one occurrence. Thus, for $M[x^B]$, $M[N^B]$ denotes the result of substituting N for occurrences of x that together count for one in $M[x]$. Let $PROOF_{ISPL_{/,\backslash,\circ}}$ denote the set of proofs in $ISPL_{/,\backslash,\circ}$. Next, we shall define a fragment of $\Lambda_{/,\backslash,\circ}$ such that every $\Pi \in PROOF_{ISPL_{/,\backslash,\circ}}$ can be encoded by some term from this fragment, and vice versa.

Definition 5.30 Let $\Lambda_{ISPL_{/,\backslash,\circ}}$ be the biggest $\Gamma \subseteq \Lambda_{/,\backslash,\circ}$ such that

(i) for every $M \in \Gamma$ and every prefix of the form $\lambda^r x$ resp. $\lambda^l x$ in M, $\lambda^r x$ resp. $\lambda^l x$ binds exactly one fvo (i.e. binds free occurrences of x that together count for one);

(ii) for every $\lambda^r x.M \in \Gamma$, an occurrence of x is the rightmost fvo in M;

(iii) for every $\lambda^l x.M \in \Gamma$, an occurrence of x is the leftmost fvo in M.

Definition 5.31 A term $M^B \in \Lambda_{ISPL_{/,\backslash,\circ}}$ is a construction of a sequent $A_1 \ldots A_n \rightarrow B$ iff M has exactly fvos $x_1^{A_1}, \ldots, x_n^{A_n}$, in this order from left to right.

Theorem 5.32 Given a proof in $PROOF_{ISPL_{/,\backslash,\circ}}$ of a sequent $s = A_1 \ldots A_n \rightarrow B$, one can find a construction $M^B \in \Lambda_{ISPL_{/,\backslash,\circ}}$ of s, and conversely.

PROOF We shall inductively define encoding functions $f_1 : PROOF_{ISPL_{/,\backslash,\circ}} \longrightarrow \Lambda_{ISPL_{/,\backslash,\circ}}$, $g_1 : \Lambda_{ISPL_{/,\backslash,\circ}} \longrightarrow PROOF_{ISPL_{/,\backslash,\circ}}$, such that it can readily be seen that $f_1(\Pi)$ is a construction of Π, and $g_1(M)$ proves a sequent of which M is a construction. Let $(< M, N >)^\flat$ denote the term which results from $< M, N >$ by renaming (from left to right) modulo counting for one the fvos in N by occurrences of *distinct* variables of the respective types and with smallest possible indices such that the new variables are distinct from the free variables occurring in M. The function f_1 is inductively defined in the same way as the function f in the proof of Theorem 5.11, except that (i) fvos that count for one are renamed by the same variable, and (ii) in addition we have:

- $\Pi = \dfrac{\dfrac{\Pi_1}{X \rightarrow A} \quad \dfrac{\Pi_2}{Y \rightarrow B}}{XY \rightarrow (A \circ B)}$: $f_1(\Pi) = (< f_1(\frac{\Pi_1}{X \rightarrow A}), f_1(\frac{\Pi_2}{Y \rightarrow B}) >^{(A \circ B)})^\flat$;
- $\Pi = \dfrac{\dfrac{\Pi_1}{XABY \rightarrow C}}{X(A \circ B)Y \rightarrow C}$: $f_1(\Pi) = (f_1(\frac{\Pi_1}{XABY \rightarrow C})[(v_i^{(A \circ B)})_0, (v_i^{(A \circ B)})_1])^\natural$,
  where v_i is the first variable of type $(A \circ B)$ not occurring in $f_1(\frac{\Pi_1}{XABY \rightarrow C})$.

The function g_1 is inductively defined in the same way as the function g in the proof of Theorem 5.11, except that in addition we have:

- $M \equiv < G, H >^{(A \circ C)}$: $g_1(M) = \dfrac{g_1(G) \quad g_1(H)}{XY \rightarrow (A \circ C)}$,
  where X resp. Y is the sequence of types of the fvos in G resp. H;
- $M \equiv N^B[(z^{(A \circ C)})_0, (z^{(A \circ C)})_1]$: $g(M) = \dfrac{g_1(N[v_i^A, v_j^C])}{X(A \circ C)Y \rightarrow B}$,
  where v_i resp. v_j is the first variable of type A resp. type C not occurring in $N[(z)_0, (z)_1]$, and $XACY$ is the sequence of types of the fvos in $N[v_i, v_j]$. □

Let $ELIM_c$ now denote our terminating cut-elimination algorithm for $ISPL_{/,\backslash,\circ}$.

Theorem 5.33 Let $\mathcal{A}_2 =< PROOF_{ISPL_{/,\backslash,\circ}}, ELIM_c >$, $\mathcal{B}_2 =< \Lambda_{ISPL_{/,\backslash,\circ}}, NORM_{\succeq} >$.

(i) The function f_1 defined in the proof of theorem 5.33 is a homomorphism from $\mathcal{A}_2$ to $\mathcal{B}_2$.

(ii) The function g_1 defined in the proof of Theorem 5.33 is a homomorphism from $\mathcal{B}_2$ to $\mathcal{A}_2$.

PROOF Note that $\Lambda_{ISPL_{/,\backslash,\circ}}$ is closed under $NORM_{\succeq}$. The proof is straightforward. To prove (i) consider the following f_1-images of proofs from the description of $ELIM_c$:

$$\left[\dfrac{\dfrac{\dfrac{\Pi_1}{X_1BCY_1\to A}}{X_1(B\circ C)Y_1\to A} \quad \dfrac{\Pi_2}{XAZ\to D}}{XX_1(B\circ C)Y_1Z\to D}\right] \text{ is converted into } \left[\dfrac{\dfrac{\dfrac{\Pi_1}{X_1BCY_1\to A} \quad \dfrac{\Pi_2}{XAZ\to D}}{XX_1BCY_1Z\to D}}{XX_1(B\circ C)Y_1Z\to D}\right]$$

$$\downarrow f_1 \qquad\qquad\qquad \downarrow f_1$$

$$\dfrac{\dfrac{f_1(\frac{\Pi_1}{X_1BCY_1\to A})}{(f_1(\frac{\Pi_1}{X_1BCY_1\to A})\,[(v_i^{(B\circ C)})_0,(v_i)_1])^{\natural}} \quad f_1(\frac{\Pi_2}{XAZ\to D})}{(N\,[(f_1(\frac{\Pi_1}{X_1BCY_1\to A})[(v_i)_0,(v_i)_1])^{\natural}])^{\natural}} \equiv \dfrac{\dfrac{f_1(\frac{\Pi_1}{X_1BCY_1\to A}) \quad f_1(\frac{\Pi_2}{XAZ\to D})}{(f_1(\frac{\Pi_2}{XAZ\to D})\,[(f_1(\frac{\Pi_1}{X_1BCY_1\to A}))^{\natural}])^{\natural}}}{((N\,[(f_1(\frac{\Pi_1}{X_1BCY_1\to A}))^{\natural}])^{\natural}[(v_i)_0,(v_i)_1])^{\natural}},$$

where $N \equiv f_1(\frac{\Pi_2}{XAZ\to D})$;

$$\left[\dfrac{\dfrac{\Pi_1}{Y\to A} \quad \dfrac{\dfrac{\Pi_2}{X_1AX_2\to B} \quad \dfrac{\Pi_3}{Z\to C}}{X_1AX_2Z\to(B\circ C)}}{X_1YX_2Z\to(B\circ C)}\right] \text{ is converted into } \left[\dfrac{\dfrac{\dfrac{\Pi_1}{Y\to A} \quad \dfrac{\Pi_2}{X_1AX_2\to B}}{X_1YX_2\to B} \quad \dfrac{\Pi_3}{Z\to C}}{X_1YX_2Z\to(B\circ C)}\right]$$

$$\downarrow f_1 \qquad\qquad\qquad \downarrow f_1$$

$$\dfrac{f_1(\frac{\Pi_1}{Y\to A}) \quad \dfrac{f_1(\frac{\Pi_2}{X_1AX_2\to B}) \quad f_1(\frac{\Pi_3}{Z\to C})}{(< f_1(\frac{\Pi_2}{X_1AX_2\to B}), f_1(\frac{\Pi_3}{Z\to C}) >)^{\flat}}}{((< f_1(\frac{\Pi_2}{X_1AX_2\to B}), f_1(\frac{\Pi_3}{Z\to C}) >)^{\flat}\,[(N)^{\natural}])^{\natural}} \equiv \dfrac{\dfrac{f_1(\frac{\Pi_1}{Y\to A}) \quad f_1(\frac{\Pi_2}{X_1AX_2\to B})}{(f_1(\frac{\Pi_2}{X_1AX_2\to B})\,[(f_1(\frac{\Pi_1}{Y\to A}))^{\natural}])^{\natural}} \quad f_1(\frac{\Pi_3}{Z\to C})}{(< (f_1(\frac{\Pi_2}{X_1AX_2\to B})\,[(N)^{\natural}])^{\natural}, f_1(\frac{\Pi_3}{Z\to C}) >)^{\flat}},$$

where $N \equiv f_1(\frac{\Pi_1}{Y\to A})$;

$$\left[\dfrac{\dfrac{\Pi_1}{Y\to A} \quad \dfrac{\dfrac{\Pi_2}{X_1AX_2BCZ\to D}}{X_1AX_2(B\circ C)Z\to D}}{X_1YX_2(B\circ C)Z\to D}\right] \text{ is converted into } \left[\dfrac{\dfrac{\dfrac{\Pi_1}{Y\to A} \quad \dfrac{\Pi_2}{X_1AX_2BCZ\to D}}{X_1YX_2BCZ\to B}}{X_1YX_2(B\circ C)Z\to B}\right]$$

$$\downarrow f_1 \qquad\qquad\qquad \downarrow f_1$$

$$\dfrac{f_1(\frac{\Pi_1}{Y\to A}) \quad \dfrac{f_1(\frac{\Pi_2}{X_1AX_2BCZ\to D})}{(f_1(\frac{\Pi_2}{X_1AX_2BCZ\to D})\,[(v_i^{(B\circ C)})_0,(v_i)_1])^{\natural}}}{((f_1(\frac{\Pi_2}{X_1AX_2BCZ\to D})\,[(v_i)_0,(v_i)_1])^{\natural}[(N)^{\natural}])^{\natural}} \equiv \dfrac{\dfrac{f_1(\frac{\Pi_1}{Y\to A}) \quad f_1(\frac{\Pi_2}{X_1AX_2BCZ\to D})}{(f_1(\frac{\Pi_2}{X_1AX_2BCZ\to D})[(f_1(\frac{\Pi_1}{Y\to A}))^{\natural}])^{\natural}}}{((f_1(\frac{\Pi_2}{X_1AX_2BCZ\to B})\,[(N)^{\natural}])^{\natural}[(v_i)_0,(v_i)_1])^{\natural}},$$

where $N \equiv f_1(\frac{\Pi_1}{Y\to A})$;

$$\left[\dfrac{\dfrac{\dfrac{\Pi_1}{X\to B}\quad\dfrac{\Pi_2}{Y\to C}}{XY\to(B\circ C)}\quad\dfrac{\dfrac{\Pi_3}{X_1BCX_2\to A}}{X_1(B\circ C)X_2\to A}}{X_1XYX_2\to A}\right]\quad\text{is converted into}\quad\left[\dfrac{\dfrac{\Pi_1}{X\to B}\quad\dfrac{\dfrac{\Pi_2}{Y\to C}\quad\dfrac{\Pi_3}{X_1BCX_2\to A}}{X_1BYX_2\to A}}{X_1XYX_2\to A}\right]$$

$$\downarrow f_1 \qquad\qquad\qquad\qquad \downarrow f_1$$

$$\dfrac{\dfrac{f_1(\frac{\Pi_1}{X\to B})\quad f_1(\frac{\Pi_2}{Y\to C})}{(< f_1(\frac{\Pi_1}{X\to B}), f_1(\frac{\Pi_2}{Y\to C}) >)^\flat}\quad\dfrac{f_1(\frac{\Pi_3}{X_1BCX_2\to A})}{(N\,[(v_i^{(B\circ C)})_0,(v_i)_1])^\natural}}{((N\,[(v_i)_0,(v_i)_1])^\natural\,[((< f_1(\frac{\Pi_1}{X\to B}), f_1(\frac{\Pi_2}{Y\to C}) >)^\flat)^\natural])^\natural}\succeq\dfrac{f_1(\frac{\Pi_1}{X\to B})\quad\dfrac{f_1(\frac{\Pi_2}{Y\to C})\quad f_1(\frac{\Pi_3}{X_1BCX_2\to A})}{(N\,[(f_1(\frac{\Pi_2}{Y\to C}))^\natural])^\natural}}{((N\,[(f_1(\frac{\Pi_2}{Y\to C}))^\natural])^\natural\,[(f_1(\frac{\Pi_1}{X\to B}))^\natural])^\natural},$$

where $N\equiv f_1(\frac{\Pi_3}{X_1BCX_2\to A})$. □

REMARK Note that the surjectivity rule $<(M)_0,(M)_1>\succ M$ covers conversions of the following form:

$$\left[\dfrac{\dfrac{A\to A\quad B\to B}{AB\to(A\circ B)}}{(A\circ B)\to(A\circ B)}\right]\text{ is converted into }\quad[(A\circ B)\to(A\circ B)].$$

5.6 Encoding proofs in structural extensions of $ISPL_{/,\backslash}$

In order to encode proofs in extensions of $ISPL_{/,\backslash}$ which are obtained by adding combinations of the structural rules **P**, **C**, **C**′, and **M**, we shall introduce for each $R\in\{\mathbf{P},\mathbf{C},\mathbf{C}',\mathbf{M}\}$ an operation $\underline{R}$ on sequences of variable occurrences, and an operation $\overline{R}$ which will be applied to the definition of the set of encoding terms.[3] This will give us the necessary flexibility, since not only the set of encoding terms but also the notion of construction has to be adjusted to enlarged sets of proofs.

EXAMPLE Consider the following application of **M**: $X\to A\vdash XB\to A$, where $X\to A$ is proved in $ISPL_{/,\backslash}$ and B is a formula in $\{/,\backslash\}$. To cover this case we may introduce a more liberal notion of construction such that a construction $M^A\in\Lambda_{ISPL_{/,\backslash}}$ of $X\to A$ is also considered a construction of $XB\to A$. If now $(\to /)$ is applied to the latter sequent, according to the earlier methodology this should correspond to a λ^r-abstraction on M^A. But this is then an empty abstraction, leading outside of $\Lambda_{ISPL_{/,\backslash}}$.

Let $\vec{x}_n^{A_n}$ abbreviate $x_1^{A_1}\ldots x_n^{A_n}$, let $\underline{\mathbf{P}}\,\vec{x}_n$ denote an arbitrary permutation of $\vec{x}_n$, let $\underline{\mathbf{C}}\,\vec{x}_n=\underline{\mathbf{C}'}\,\vec{x}_n$ denote just $\vec{x}_n$, and let $\underline{\mathbf{M}}\,\vec{x}_n$ denote the result of deleting some (possibly all) occurrences in $\vec{x}_n$. Let $M(\vec{x}_n)$ denote a term with the sequence $\vec{x}_n$ of fvos, and let for $M(\vec{x}_n\;y_1^Ay_2^A\;\vec{z}_m)$ resp. $M(\vec{x}_n\;y_1^A\;\vec{z}_m\;y_2^A\;\vec{w}_l)$, $M^\bullet$ resp. M° denote $M(\vec{x}_n\;v_i^Av_i^A\;\vec{z}_m)$ resp. $M(\vec{x}_n\;v_i^A\;\vec{z}_m\;v_i^A\;\vec{w}_l)$, where v_i is the first variable of type A not occurring in $M(\vec{x}_n\;y_1y_2\;\vec{z}_m)$ resp. $M(\vec{x}_n\;y_1\;\vec{z}_m\;y_2\;\vec{w}_l)$, and the occurrences of v_i are said to count for one in $M^\bullet$ resp. M°. In the present and in the following section, renaming of fvos

[3] The rules **E**, **E**′ are excluded from these considerations for reasons which will be explained later on.

and substitution for single occurrences is to be understood wrt this notion of counting for one. In particular, for $M[x^A]$, distinct fvos counting for one in G^A count for one in $M[G]$.

Definition 5.34 For $R \in \{\mathbf{P}, \mathbf{C}, \mathbf{C'}, \mathbf{M}\}$, the operation $\overline{R}$ on the definition of $\Lambda_{ISPL_{/,\backslash}}$ is defined as follows:

$\overline{\mathbf{P}}$: drop clauses (ii) and (iii);

$\overline{\mathbf{C}}$: add to clause (i) "or $\lambda^r x$ resp. $\lambda^l x$ binds more than one fvo, provided these occurrences are adjacent in M's sequence of fvos";

$\overline{\mathbf{C'}}$: add to clause (i) "or $\lambda^r x$ resp. $\lambda^l x$ binds at least one fvo";

$\overline{\mathbf{M}}$: add to clause (i) "or $\lambda^r x$ resp. $\lambda^l x$ binds at most one fvo".

Let $\Theta \subseteq \{\mathbf{P}, \mathbf{C}, \mathbf{C'}, \mathbf{M}\}$.

Definition 5.35 Let $\overline{\Theta} = \{\overline{R} \mid R \in \Theta\}$. $\Lambda_{ISPL_{/,\backslash\Theta}}$ is defined by successively applying every $\overline{R} \in \overline{\Theta}$ to definition 3.1 and at the same time replacing "$\Lambda_{ISPL_{/,\backslash}}$" by "$\Lambda_{ISPL_{/,\backslash\Theta}}$".

Definition 5.36 Let $\underline{\Theta} = \{\underline{R} \mid R \in \Theta\}$. $M^B \in \Lambda_{ISPL_{/,\backslash\Theta}}$ is a Θ-construction of a sequent $A_1 \dots A_n \rightarrow B$ iff M's sequence of fvos is the result of applying a finite combination of $\underline{R}$'s $\in \underline{\Theta}$ to a sequence of occurrences $x_1^{A_1} \dots x_n^{A_n}$.[4]

Theorem 5.37 Given a proof in $ISPL_{/,\backslash\Theta}$ of a sequent $s = A_1 \dots A_n \rightarrow B$, one can find a Θ-construction $M^B \in \Lambda_{ISPL_{/,\backslash\Theta}}$ of s, and conversely.

PROOF Let $PROOF_{ISPL_{/,\backslash\Theta}}$ denote the set of proofs in $ISPL_{/,\backslash\Theta}$. We shall define functions $f^\Theta : PROOF_{ISPL_{/,\backslash\Theta}} \longrightarrow \Lambda_{ISPL_{/,\backslash\Theta}}$, $g^\Theta : \Lambda_{ISPL_{/,\backslash\Theta}} \longrightarrow PROOF_{ISPL_{/,\backslash\Theta}}$, such that it can easily be verified that $f^\Theta(\Pi)$ is a Θ-construction of Π, and $g^\Theta(M)$ proves a sequent of which M is a Θ-construction. The function f^Θ is inductively defined as follows:

- $\Pi = A \rightarrow A$: $f^\Theta(\Pi) \equiv v_1^A$.

- $\Pi = \frac{\frac{\Pi_1}{XA \rightarrow B}}{X \rightarrow (B/A)}$: $f^\Theta(\Pi) \equiv \lambda^r v_i^A . f^\Theta(\frac{\Pi_1}{XA \rightarrow B})$,
  where an occurrence of v_i is the rightmost fvo of type A in $f^\Theta(\frac{\Pi_1}{XA \rightarrow B})$, provided there is a fvo of type A in $f^\Theta(\frac{\Pi_1}{XA \rightarrow B})$;
  where v_i is the first variable of type A not occurring in $f^\Theta(\frac{\Pi_1}{XA \rightarrow B})$, otherwise.

- $\Pi = \frac{\frac{\Pi_1}{Y \rightarrow A} \quad \frac{\Pi_2}{XBZ \rightarrow C}}{X(B/A)YZ \rightarrow C}$: Case 1: There is an occurrence of B in the sequence of types of $f^\Theta(\frac{\Pi_2}{XBZ \rightarrow C})$'s sequence of fvos. $f^\Theta(\Pi) \equiv (f^\Theta(\frac{\Pi_2}{XBZ \rightarrow C})[(v_j^{(B/A)} f^\Theta(\frac{\Pi_1}{Y \rightarrow A}))^\natural])^\natural$,
  where v_j is the first variable of type (B/A) not occurring in $f^\Theta(\frac{\Pi_1}{Y \rightarrow A})$.
  Case 2: There is no occurrence of B in the sequence of types of $f^\Theta(\frac{\Pi_2}{XBZ \rightarrow C})$'s sequence of fvos. $f^\Theta(\Pi) \equiv f^\Theta(\frac{\Pi_2}{XBZ \rightarrow C})$.

[4] Thus, an $\emptyset$-construction is a construction in the original sense.

- for $\Pi = \frac{\frac{\Pi_1}{AX\to B}}{X\to(A\backslash B)}$ resp. $\Pi = \frac{\frac{\Pi_1}{Y\to A}\quad\frac{\Pi_2}{XBZ\to C}}{XY(A\backslash B)Z\to C}$, $f^{\ominus}(\Pi)$ is defined in analogy to the previous two cases.

- $\Pi = \frac{\frac{\Pi_1}{Y\to A}\quad\frac{\Pi_2}{XAZ\to B}}{XYZ\to B}$: Case 1: There is an occurrence of A in the sequence of types of $f^{\ominus}(\frac{\Pi_2}{XAZ\to B})$'s sequence of fvos. $f^{\ominus}(\Pi) = (f^{\ominus}(\frac{\Pi_2}{XAZ\to B})\,[(f^{\ominus}(\frac{\Pi_1}{Y\to A}))^{\natural}])^{\natural}$.
  Case 2: There is no occurrence of A in the sequence of types of $f^{\ominus}(\frac{\Pi_2}{XAZ\to B})$'s sequence of fvos. $f^{\ominus}(\Pi) \equiv f^{\ominus}(\frac{\Pi_2}{XAZ\to B})$.

- Π is **P** or **M** applied to $\frac{\Pi_1}{X\to A}$: $f^{\ominus}(\Pi) \equiv f^{\ominus}(\frac{\Pi_1}{X\to A})$.

- Π is **C** resp. **C′** applied to $\frac{\Pi_1}{XBBY\to A}$ resp. $\frac{\Pi_1}{XBYBZ\to A}$: $f^{\ominus}(\Pi) \equiv ((f^{\ominus}(\frac{\Pi_1}{XBBY\to A}))^{*})^{\natural}$ resp. $f^{\ominus}(\Pi) \equiv ((f^{\ominus}(\frac{\Pi_1}{XBYBZ\to A}))^{\diamond})^{\natural}$.

The function $g^{\ominus}$ is inductively defined by using the clauses from the definition of g in the proof of Theorem 5.11, replacing "g" by "$g^{\ominus}$", except that we now have:

- $M \equiv \lambda^{r}x^{A}.N^{C}$, where X is the sequence of types of N's sequence of fvos:
  Case 1: $x^{A} \in FV(N)$. $g^{\ominus}(M) = \frac{g^{\ominus}(N)}{X[-A]\to(C/A)}$, where $X[-A]$ is the result of removing from X those occurrences of A which correspond to occurrences of x in N;
  Case 2: $x^{A} \notin FV(N)$. $g^{\ominus}(M) = \frac{\frac{g^{\ominus}(N)}{XA\to C}}{X\to(C/A)}$.

- $M^{B} \equiv \lambda^{l}x^{A}.N^{C}$: analogous to the previous case. □

Note that in general it is not true that $g^{\ominus}(f^{\ominus}(\Pi)) = \Pi$. The rules **E**, **E′** have been excluded from the above considerations for the following reasons: Suppose one wants to define a notion of construction such that every construction $M^{B}(\vec{x}_{n}^{\vec{A}_n})$ of the premise $A_1 \ldots A_n \to B$ in an application of **E** or **E′** is already a construction of the conclusion of this application. In the case of **E** one might start with requiring that the sequence of M's fvos is the result of deleting occurrences in $\vec{x}_n$ which repeat some adjacent occurrence of the same type. This does not, however, give an appropriate notion of construction, because the construction property may be spoiled by applying (*cut*) after applying **E**. In the case of **E′** additional difficulties arise concerning the order of fvos.

It would be a bit tedious, though not impossible, to extend the previous theorem to $ISPL_{/,\backslash,\circ\ominus}$. As far as generalizations are concerned, the *really* interesting problem is to extend the formulas-as-types approach to the 'additive' connectives $\wedge$ and $\vee$. Howard's [1969] method of 'closed prime terms' is very ad hoc. A more natural and uniform strategy is introducing type-forming operations corresponding to $\wedge$ and $\vee$, as suggested by van Benthem [1991, Chapters 2 and 11] in his Boolean Lambda Calculus.

5.7 Cut-elimination in $PROOF_{ISPL_{/,\backslash\ominus}}$ and β-reduction in $\Lambda_{ISPL_{/,\backslash\ominus}}$ as homomorphic images of each other

Theorem 5.38 Let $ELIM_{c}^{\ominus}$ denote the cut-elimination algorithm for $PROOF_{ISPL_{/,\backslash\ominus}}$.

(i) $f^{\ominus}$ is a homomorphism from $< PROOF_{ISPL_{/,\backslash\ominus}}, ELIM_c^{\ominus} >$ to $< \Lambda_{ISPL_{/,\backslash\ominus}}, NORM_\beta >$.
(ii) $g^{\ominus}$ is a homomorphism from $< \Lambda_{ISPL_{/,\backslash\ominus}}, NORM_\beta >$ to $< PROOF_{ISPL_{/,\backslash\ominus}}, ELIM_c^{\ominus} >$.

Proof Note that the respective sets of terms are closed under β-reduction, consider the following $f^{\ominus}$-images of proofs from the description of $ELIM_c^{\ominus}$, and check the homomorphism property:
$\mathsf{R} = \mathbf{P}$ or $\mathbf{M}$:

$$\left[\frac{\dfrac{\dfrac{\Pi_1}{Y_1\to A}}{Y_2\to A}\,\mathsf{R} \qquad \dfrac{\Pi_2}{Z_1AZ_2\to B}}{Z_1Y_2Z_2\to B}\right] \text{ is converted into } \left[\dfrac{\dfrac{\dfrac{\Pi_1}{Y_1\to A}\quad \dfrac{\Pi_2}{Z_1AZ_2\to B}}{Z_1Y_1Z_2\to B}}{Z_1Y_2Z_2\to B}\,\mathsf{R}\right]$$

$$\downarrow f^{\ominus} \qquad\qquad \downarrow f^{\ominus}$$

$$\frac{\dfrac{f^{\ominus}(\frac{\Pi_1}{Y_1\to A})}{f^{\ominus}(\frac{\Pi_1}{Y_1\to A})}\quad f^{\ominus}(\frac{\Pi_2}{Z_1AZ_2\to B})}{(f^{\ominus}(\frac{\Pi_2}{Z_1AZ_2\to B})[(f^{\ominus}(\frac{\Pi_1}{Y_1\to A}))^{\natural}])^{\natural}} \equiv \frac{\dfrac{f^{\ominus}(\frac{\Pi_1}{Y_1\to A})\quad f^{\ominus}(\frac{\Pi_2}{Z_1AZ_2\to B})}{(f^{\ominus}(\frac{\Pi_2}{Z_1AZ_2\to B})[(f^{\ominus}(\frac{\Pi_1}{Y_1\to A}))^{\natural}])^{\natural}}}{(f^{\ominus}(\frac{\Pi_2}{Z_1AZ_2\to B})[(f^{\ominus}(\frac{\Pi_1}{Y_1\to A}))^{\natural}])^{\natural}};$$

$\mathsf{R} = \mathbf{C}$:

$$\left[\frac{\dfrac{\dfrac{\Pi_1}{Y_1\to A}}{Y_2\to A}\,\mathsf{R} \qquad \dfrac{\Pi_2}{Z_1AZ_2\to B}}{Z_1Y_2Z_2\to B}\right] \text{ is converted into } \left[\dfrac{\dfrac{\dfrac{\Pi_1}{Y_1\to A}\quad \dfrac{\Pi_2}{Z_1AZ_2\to B}}{Z_1Y_1Z_2\to B}}{Z_1Y_2Z_2\to B}\,\mathsf{R}\right]$$

$$\downarrow f^{\ominus} \qquad\qquad \downarrow f^{\ominus}$$

$$\frac{\dfrac{f^{\ominus}(\frac{\Pi_1}{Y_1\to A})}{((f^{\ominus}(\frac{\Pi_1}{Y_1\to A}))^{*})^{\natural}}\quad f^{\ominus}(\frac{\Pi_2}{Z_1AZ_2\to B})}{(f^{\ominus}(\frac{\Pi_2}{Z_1AZ_2\to B})[((f^{\ominus}(\frac{\Pi_1}{Y_1\to A}))^{*})^{\natural}])^{\natural}} \equiv \frac{\dfrac{f^{\ominus}(\frac{\Pi_1}{Y_1\to A})\quad f^{\ominus}(\frac{\Pi_2}{Z_1AZ_2\to B})}{(f^{\ominus}(\frac{\Pi_2}{Z_1AZ_2\to B})[(f^{\ominus}(\frac{\Pi_1}{Y_1\to A}))^{\natural}])^{\natural}}}{(((f^{\ominus}(\frac{\Pi_2}{Z_1AZ_2\to B})[(f^{\ominus}(\frac{\Pi_1}{Y_1\to A}))^{\natural}])^{\natural})^{*})^{\natural}};$$

$$\left[\frac{\dfrac{\Pi_1}{Y_1\to A} \qquad \dfrac{\dfrac{\Pi_2}{Z_1AZ_2\to B}}{Z_3AZ_4\to B}\,\mathsf{R}}{Z_3Y_1Z_4\to B}\right] \text{ is converted into } \left[\dfrac{\dfrac{\dfrac{\Pi_1}{Y_1\to A}\quad \dfrac{\Pi_2}{Z_1AZ_2\to B}}{Z_1Y_1Z_2\to B}}{Z_3Y_1Z_4\to B}\,\mathsf{R}\right]$$

$$\downarrow f^{\ominus} \qquad\qquad \downarrow f^{\ominus}$$

$$\frac{f^{\ominus}(\frac{\Pi_1}{Y_1\to A})\quad \dfrac{f^{\ominus}(\frac{\Pi_2}{Z_1AZ_2\to B})}{((f^{\ominus}(\frac{\Pi_2}{Z_1AZ_2\to B}))^{*})^{\natural}}}{(f^{\ominus}(\frac{\Pi_2}{Z_1AZ_2\to B})[((f^{\ominus}(\frac{\Pi_1}{Y_1\to A}))^{*})^{\natural}])^{\natural}} \equiv \frac{\dfrac{f^{\ominus}(\frac{\Pi_1}{Y_1\to A})\quad f^{\ominus}(\frac{\Pi_2}{Z_1AZ_2\to B})}{(f^{\ominus}(\frac{\Pi_2}{Z_1AZ_2\to B})[(f^{\ominus}(\frac{\Pi_1}{Y_1\to A}))^{\natural}])^{\natural}}}{(((f^{\ominus}(\frac{\Pi_2}{Z_1AZ_2\to B})[(f^{\ominus}(\frac{\Pi_1}{Y_1\to A}))^{\natural}])^{\natural})^{*})^{\natural}};$$

$$\left[\dfrac{\dfrac{\Pi_1}{Y_1\to A}\qquad \dfrac{\dfrac{\Pi_2}{Z_1AZ_2AZ_3\to B}}{Z_1Z_2AZ_3\to B}}{Z_1Z_2Y_1Z_3\to B}\right]\quad\text{is converted into}\quad\left[\dfrac{\dfrac{\dfrac{\Pi_1}{Y_1\to A}\qquad\dfrac{\dfrac{\Pi_1}{Y_1\to A}\quad\dfrac{\Pi_2}{Z_1AZ_2AZ_3\to B}}{Z_1Y_1Z_2AZ_3\to B}}{Z_1Y_1Z_2Y_1Z_3\to B}}{\dfrac{\vdots}{Z_1Z_2Y_1Z_3\to B}}\right]$$

$$\downarrow f^{\Theta}\qquad\qquad\qquad\qquad\downarrow f^{\Theta}$$

$$\dfrac{f^{\Theta}(\frac{\Pi_1}{Y_1\to A})\quad \dfrac{\underline{N}}{((N)^{\circ})^{\natural}}}{(((N)^{\circ})^{\natural}[(f(\frac{\Pi_1}{Y_1\to A}))^{\natural}])^{\natural}}\ \equiv\ \dfrac{\dfrac{f^{\Theta}(\frac{\Pi_1}{Y_1\to A})\quad\dfrac{f^{\Theta}(\frac{\Pi_1}{Y_1\to A})\quad N}{(N[(f^{\Theta}(\frac{\Pi_1}{Y_1\to A}))^{\natural}])^{\natural}}}{((N[(f^{\Theta}(\frac{\Pi_1}{Y_1\to A}))^{\natural}])^{\natural}[(f^{\Theta}(\frac{\Pi_1}{Y_1\to A}))^{\natural}])^{\natural}}}{\dfrac{\vdots}{((\ldots((((N[(f^{\Theta}(\frac{\Pi_1}{Y_1\to A}))^{\natural}])^{\natural}[(f^{\Theta}(\frac{\Pi_1}{Y_1\to A}))^{\natural}])^{\natural})^{\circ})^{\natural}\ldots)^{\circ})^{\natural}}},$$

where $N \equiv f^{\Theta}(\frac{\Pi_2}{Z_1AZ_2AZ_3\to B})$;

$$\left[\dfrac{\dfrac{\Pi_1}{Y_1\to A}\qquad\dfrac{\dfrac{\Pi_2}{Z_1Z_2\to B}}{Z_1AZ_2\to B}}{Z_1Y_1Z_2\to B}\right]\quad\text{is converted into}\quad\left[\dfrac{\dfrac{\Pi_2}{Z_1Z_2\to B}}{\dfrac{\vdots}{Z_1Y_1Z_2\to B}}\right]$$

$$\downarrow f^{\Theta}\qquad\qquad\qquad\qquad\downarrow f^{\Theta}$$

$$\dfrac{f^{\Theta}(\frac{\Pi_1}{Y_1\to A})\quad\dfrac{f^{\Theta}(\frac{\Pi_2}{Z_1Z_2\to B})}{f^{\Theta}(\frac{\Pi_2}{Z_1Z_2\to B})}}{f^{\Theta}(\frac{\Pi_2}{Z_1Z_2\to B})}\ \equiv\ \dfrac{f^{\Theta}(\frac{\Pi_2}{Z_1Z_2\to B})}{\dfrac{\vdots}{f^{\Theta}(\frac{\Pi_2}{Z_1Z_2\to B})}}.\quad\square$$

Equality of terms in the ordinary typed lambda calculus $\lambda_{\supset}$ enjoys a very natural set-theoretic characterization by so-called 'full type structures over infinite sets' (see [Friedman 1975] and Chapter 8 below). In these structures, for every propositional variable p there is an infinite domain D^p providing denotations for terms of type p. Domains having in store denotations for terms of an implicational type $(A \supset B)$ are defined as the sets of all functions from D^A to D^B, i.e. $D^{(A\supset B)} \stackrel{def}{=} (D^B)^{D^A}$. In this way, proofs in $ISPL_{\supset}$ receive *denotations* via their encoding by means of typed λ-terms. Van Benthem [1986, 1991] refers to these terms as the readings or *meanings* of the proofs they encode. Since there is no mathematical notion of 'directional function' the usual set-theoretic interpretation is not available for terms of a directional type (A/B) resp. $(B \setminus A)$ (at least if these terms encode proofs in sequent calculi without **P**, which would not allow the replacement of λ^r, λ^l resp. $/$, $\setminus$ by λ resp. $\supset$). This does, of course, *not* mean that terms of a directional type cannot be interpreted in illuminating ways. Interesting natural candidates e.g. are provided by van Benthem's [1991] L-models. We can show that for any adequate assignment of denotations to $T_{/,\setminus}$-terms,

Theorem 5.39 If $X \to A$ is provable in $ISPL_{/,\backslash\Delta_1}$, $\Delta_1 \subseteq \{\mathbf{P}, \mathbf{M}\}$, then, up to β-nf's, there are only finitely many Δ_1-constructions of this sequent in $\Lambda_{/,\backslash\Delta_1}$.

PROOF By a straightforward extension of an argument from [van Benthem 1991, p. 115]. As we have seen, cut-elimination steps in $PROOF_{ISPL_{/,\backslash\Delta_1}}$ do not affect the encoding terms, i.e. meanings, up to β-nf's. Therefore the proofs encoded by all possible meanings up to β-nf's are among the (*cut*)-free proofs. Now, since complete proof-search trees for (*cut*)-free proofs in $ISPL_{/,\backslash\Delta_1}$ are finite, the number of meanings is also finite. □

5.8 Appendix: Proof of strong normalization for $\Lambda_{/,\backslash}$ wrt $\twoheadrightarrow_\beta$ and $\twoheadrightarrow_{\beta\eta}$

Proof of Theorem 5.8 This theorem can be proved in the same way as the corresponding result for $\lambda_\supset$ in [Hindley & Seldin 1986, Appendix 2]. The theorem follows from two lemmas; to prove the second lemma yet another preparatory lemma is required.

Definition 5.40 The set of strongly computable (sc) $T_{/,\backslash}$-terms is the smallest set Γ such that

(i) $M^p \in \Gamma$ iff M^p is sn;

(ii) $M^{(A\backslash B)} \in \Gamma$ iff for every N^A, NM is sc;

(iii) $M^{(B/A)} \in \Gamma$ iff for every N^A, MN is sc.

We shall use the following abbreviations:

$$\vec{M}_n^{A_n}\ N \equiv \begin{cases} (\ldots(M_1^{A_1}(\ldots(M_n^{A_n}N)\ldots) & \text{if } n \geq 1 \\ N & \text{if } n = 0; \end{cases}$$

$$N\ \overleftarrow{M}_n^{A_n} \equiv \begin{cases} (\ldots(NM_n^{A_n})\ldots)M_1^{A_1})\ldots) & \text{if } n \geq 1 \\ N & \text{if } n = 0; \end{cases}$$

$$(\vec{A}_n \backslash B) = \begin{cases} (A_n \backslash (\ldots \backslash (A_1 \backslash B)\ldots) & \text{if } n \geq 1 \\ B & \text{if } n = 0; \end{cases}$$

$$(B/\overleftarrow{A}_n) = \begin{cases} (\ldots(B/A_1)/\ldots)/A_n) & \text{if } n \geq 1 \\ B & \text{if } n = 0. \end{cases}$$

Note that each implicational L-formula is of the form

(I) $\vec{A}_{n_j j} \backslash((\ldots \backslash((\vec{A}_{n_1 1} \backslash(p/\overleftarrow{B}_{m_1 1}))/\ldots))/\overleftarrow{B}_{m_k k})$ or

(II) $(\vec{A}_{n_j j} \backslash((\ldots \backslash((\vec{A}_{n_1 1} \backslash p)/\overleftarrow{B}_{m_1 1}))/\ldots))/\overleftarrow{B}_{m_k k}$ or

(III) $\vec{A}_{n_j j} \backslash((\ldots \backslash((\vec{A}_{n_1 1} \backslash((p/\overleftarrow{B}_{m_1 1})/\overleftarrow{B}_{m_2 2}))/\ldots))/\overleftarrow{B}_{m_k k})$ or

(IV) $(\overrightarrow{A}_{n_j j} \backslash ((\ldots \backslash ((\overleftarrow{A}_{n_2 2} \backslash (\overrightarrow{A}_{n_1 1} \backslash p))/ \overleftarrow{B}_{m_1 1}))/ \ldots))/ \overleftarrow{B}_{m_k k}$.

This can easily be verified. Every propositional variable is of the form (I), (II), (III), and (IV). Next one has to distinguish among 32 cases which come out as follows: for every $\gamma, \delta \in \{(I), (II), (III), (IV)\}$, if A is of the form (γ/δ), then A is of the form γ; if A is of the form $(\gamma \backslash \delta)$, then A is of the form δ.

REMARK (i) By the definition of strong computability, we have that

if A is of the form (I), then G^A is sc iff for all sc $M_{1_1}^{A_{1_1}}, \ldots, M_{n_1 1}^{A_{n_1 1}}, \ldots, M_{1_j}^{A_{1_j}}, \ldots, M_{n_j j}^{A_{n_j j}}, N_{m_k k}^{B_{m_k k}}, \ldots, N_{1_k}^{B_{1_k}}, \ldots, N_{m_1 1}^{B_{m_1 1}}, \ldots, N_{1_1}^{A_{1_1}}$,

$H_{(I)}^p \equiv (\overrightarrow{M}_{n_1 1} \ldots \overrightarrow{M}_{n_j j}\, G\, \overleftarrow{N}_{m_k k} \ldots \overleftarrow{N}_{m_1 1})^p$ is sn.

Analogously, $H_{(II)}^p$, $H_{(III)}^p$ and $H_{(IV)}^p$ are sn.

(ii) From the definition of strong computability it is clear that if $M^{(A \backslash B)}, N^{(B/A)}, G^A$ are sc, then GM, NG are sc.
(iii) If M is sc, then every subterm of M is sc.

Lemma 5.41 For every implicational L-formula B:

(i) if A is of the form $(\underline{I})$ $\overrightarrow{A}_{n_j j} \backslash ((\ldots \backslash ((\overrightarrow{A}_{n_1 1} \backslash (B/ \overleftarrow{B}_{m_1 1}))/ \ldots))/ \overleftarrow{B}_{m_k k})$ and $M_{1_1}^{A_{1_1}}, \ldots, M_{n_1 1}^{A_{n_1 1}}, \ldots, M_{1_j}^{A_{1_j}}, \ldots, M_{n_j j}^{A_{n_j j}}, N_{m_k k}^{B_{m_k k}}, \ldots, N_{1_k}^{B_{1_k}}, \ldots N_{m_1 1}^{B_{m_1 1}}, \ldots, N_{1_1}^{A_{1_1}}$ are sn, then $H_{(\underline{I})}^B \equiv (\overrightarrow{M}_{n_1 1} \ldots \overrightarrow{M}_{n_j j}\, x^A\, \overleftarrow{N}_{m_k k} \ldots \overleftarrow{N}_{m_1 1})^B$ is sc.
Analogously, $H_{(\underline{II})}^B$, $H_{(\underline{III})}^B$ and $H_{(\underline{IV})}^B$ are sc.

(ii) if M^B is sc, then it is sn.

PROOF By induction on the construction of B. Case 1: B is a propositional variable. (i): Since the M_i, N_l are sn, $H_{(\underline{I})}$, $H_{(\underline{II})}$, $H_{(\underline{III})}$, and $H_{(\underline{IV})}$ must be sn. By the definition of strong computability, $H_{(\underline{I})}$, $H_{(\underline{II})}$, $H_{(\underline{III})}$, and $H_{(\underline{IV})}$ are sc. (ii): By the definition of strong computability. Case 2: B is of the form $(C \backslash D)$. (i): Suppose that G^C is sc. By the induction hypothesis (ii), G is sn. The induction hypothesis (i) gives that $GH_{(\underline{I})}$, $GH_{(\underline{II})}$, $GH_{(\underline{III})}$, and $GH_{(\underline{IV})}$ are sc. By the definition of strong computability, $H_{(\underline{I})}$, $H_{(\underline{II})}$, $H_{(\underline{III})}$, and $H_{(\underline{IV})}$ are sc. (ii): Suppose that M^B is sc and that y^C does not occur in M. By the induction hypothesis (i), with $j = k = 0$, y is sc. By the above remark (ii), yM is sc. By the induction hypothesis (ii), yM is sn. But then, by the remark (iii), also M is sn. Case 3: B is of the form (D/C). This case is analogous to Case 2. □

Lemma 5.42 If $M^A[x^B := N^B]$ is sc, so are $N(\lambda^l x^B.M), (\lambda^r x^B.M)N$, provided that N is sc if x is not free in M.

PROOF Suppose that A is of the form (I), (II), (III), or (IV) and that $M_{1_1}^{A_{1_1}}, \ldots, M_{n_1 1}^{A_{n_1 1}}, \ldots, M_{1_j}^{A_{1_j}}, \ldots, M_{n_j j}^{A_{n_j j}}, N_{m_k k}^{B_{m_k k}}, \ldots, N_{1_k}^{B_{1_k}}, \ldots, N_{m_1 1}^{B_{m_1 1}}, \ldots, M_{1_1}^{A_{1_1}}$ are sc. Since $M[x := N]$ is sc, by the above remark (ii),

(a_I) $(\vec{M}_{n_1 1} \ldots \vec{M}_{n_j j}\, M^A\,[x^B := N^B]\, \overleftarrow{N}_{m_k k} \ldots \overleftarrow{N}_{m_1 1})^p$ is sn.

Analogously, (a_{II}), (a_{III}), and (a_{IV}) are sn. In view of the above remark (i) it will suffice to show that

(a_I^l) $(\vec{M}_{n_1 1} \ldots \vec{M}_{n_j j}\, N(\lambda^l x.M)\, \overleftarrow{N}_{m_k k} \ldots \overleftarrow{N}_{m_1 1})^p$,

(a_I^r) $(\vec{M}_{n_1 1} \ldots \vec{M}_{n_{j-1} j-1}\, (\lambda^r x.M)N\, \overleftarrow{N}_{m_k k}) \ldots \overleftarrow{N}_{m_1 1})^p$,

and the analogously defined $(a_{II}^l), (a_{II}^r), (a_{III}^l), (a_{III}^r), (a_{IV}^l)$, and (a_{IV}^r) are sn. Because (a_I) is sn, also all subterms of (a_I) are sn, in particular $M\,[x := N]$, M_{1^1}, ..., $M_{n_1 1}$, ..., M_{1^j}, ..., $M_{n_j j}$, $N_{m_k k}$, ..., N_{1^k}, ..., $N_{m_1 1}$, ..., N_{1^1} are sn. By hypothesis and the previous lemma, N is sn if it does not occur in $M\,[x := N]$. For this reason, an infinite β-reduction starting at (a_I^l) cannot consist completely of β-reductions in M, N, M_{1^1}, ..., $M_{n_1 1}$, ..., M_{1^j}, ..., $M_{n_j j}$, $N_{m_k k}$, ..., N_{1^k}, ..., $N_{m_1 1}$, ..., M_{1^1}. An infinite β-reduction starting at (a_I^l) must therefore have the form

$(\vec{M}_{n_1 1} \ldots \vec{M}_{n_j j}\, N(\lambda^l x.M)\, \overleftarrow{N}_{m_k k} \ldots \overleftarrow{N}_{m_1 1})^p$

$\twoheadrightarrow_\beta \vec{M'}_{n_1 1} \ldots \vec{M'}_{n_j j}\, N'(\lambda^l x.M')\, \overleftarrow{N'}_{m_k k} \ldots \overleftarrow{N'}_{m_1 1}$

(where $M_i \twoheadrightarrow_\beta M_i'$, $N_l \twoheadrightarrow_\beta N_l'$, $M \twoheadrightarrow_\beta M'$, $N \twoheadrightarrow_\beta N'$)

$\rightarrow_\beta \vec{M'}_{n_1 1} \ldots \vec{M'}_{n_j j}\, M'\,[x := N']\, \overleftarrow{N'}_{m_{k-1} k} \ldots \overleftarrow{N'}_{m_1 1} \twoheadrightarrow_\beta \ldots$.

The reductions $M \twoheadrightarrow_\beta M', N \twoheadrightarrow_\beta N'$ give $M\,[x := N] \twoheadrightarrow_\beta M'\,[x := N']$. Thus, one may construct an infinite β-reduction from (a_I):

$(\vec{M}_{n_1 1} \ldots \vec{M}_{n_j j}\, M^A\,[x := N]\, \overleftarrow{N}_{m_k k} \ldots \overleftarrow{N}_{m_1 1})^p \twoheadrightarrow_\beta$

$\vec{M'}_{n_1 1} \ldots \vec{M'}_{n_j j}\, M'\,[x := N'])\, \overleftarrow{N'}_{m_k k} \ldots \overleftarrow{N'}_{m_1 1} \twoheadrightarrow_\beta \ldots$

contradicting the fact that (a_I) is sn. Therefore (a_I^l) must be sn. Analogous reasoning applies to the remaining cases. □.

Lemma 5.43 For every $M, x_1^{A_1}, \ldots, x_n^{A_n}$ and every sc $N_1^{A_1}, \ldots, N_n^{A_n}$: $M_\star \equiv M\,[x_1 := N_1] \ldots [x_n := N_n]$ is sc (i.e., in particular, M is sc).

PROOF By induction on the construction of M. Case 1: M is a variable x_i. Then $M_\star \equiv N_i$ is sc by hypothesis. Case 2: M is a variable distinct from the x_i's. Then $M_\star \equiv M$ is sc by Lemma 5.41. Case 3: $M \equiv M_1^A M_2^{(A\backslash B)}$. Then $M_\star \equiv M_{1\star} M_{2\star}$. $M_{1\star}, M_{2\star}$ are sc by the induction hypothesis. By the remark (ii), then also $M_\star$ is sc. Case 4: $M \equiv M_2^{(B/A)} M_1^A$: analogous to Case 3. Case 5: $M \equiv \lambda^l x^A.F$. Then $M_\star \equiv \lambda^l x.F_\star$. In order to prove that $M_\star$ is sc, one must show that for every sc N^A: $NM_\star$ is sc. Now, $NM_\star \twoheadrightarrow_\beta M_\star\,[x := N]$. The latter term is sc by the induction hypothesis for the case of $n+1$ instead of n. But then, by Lemma 5.42, $NM_\star$ is sc. Case 6: $M \equiv \lambda^r x^A.F$: analogous to Case 5. □.

Now, by Lemma 5.43, M is sc, and by Lemma 5.41, it is also sn. □

Proof of Theorem 5.20 Cf. again [Hindley & Seldin 1986, Appendix 2]. The proof is the same as the proof of Theorem 5.8, except that in addition one has to take into account in the proof of Lemma 5.42 the possibility that an infinite $\beta\eta$-reduction starting at (a_I^l) may have the form

$$(\vec{M}_{n_1 1} \ldots \vec{M}_{n_j j}\ N(\lambda^l x.M)\ \overleftarrow{N}_{m_k k} \ldots \overleftarrow{N}_{m_1 1})^p$$

$$\twoheadrightarrow_{\beta\eta} \vec{M'}_{n_1 1} \ldots \vec{M'}_{n_j j}\ N'(\lambda^l x.M')\ \overleftarrow{N'}_{m_k k} \ldots \overleftarrow{N'}_{m_1 1}$$

$$\equiv \vec{M'}_{n_1 1} \ldots \vec{M'}_{n_j j}\ N'^B(\lambda^l x^B.x^B H^{(B \backslash A)})\ \overleftarrow{N'}_{m_k k} \ldots \overleftarrow{N'}_{m_1 1},\ x \notin FV(H)$$

$$\rightarrow_{\eta} \vec{M'}_{n_1 1} \ldots \vec{M'}_{n_j j}\ N'H\ \overleftarrow{N'}_{m_{k-1} k} \ldots \overleftarrow{N'}_{m_1 1} \twoheadrightarrow_{\beta\eta} \ldots .$$

But then an infinite $\beta\eta$-reduction starting at (a_I) can be constructed as follows:

$$\vec{M}_{n_1 1} \ldots \vec{M}_{n_j j}\ M\,[x := N]\ \overleftarrow{N}_{m_k k} \ldots \overleftarrow{N}_{m_1 1} \twoheadrightarrow_{\beta\eta}$$

$$\vec{M'}_{n_1 1} \ldots \vec{M'}_{n_j j}\ M'\,[x := N']\ \overleftarrow{N'}_{m_k k}) \ldots \overleftarrow{N'}_{m_1 1} \equiv$$

$$\vec{M'}_{n_1 1} \ldots \vec{M'}_{n_j j}\ N'H\ \overleftarrow{N'}_{m_k k} \ldots \overleftarrow{N'}_{m_1 1} \twoheadrightarrow_{\beta\eta} \ldots . \ \square$$

Chapter 6

Constructive minimal and constructive information processing

In Chapter 2, the idea of taking negative information seriously led us to considering Nelson's strong, constructive negation. It is the aim of the present chapter to first of all define substructural subsystems of the (propositional part) of Nelson's constructive logics $\mathbf{N}^-$ and $\mathbf{N}$ as presented in Chapter 2. The problem is to incorporate the primitive strong negation operation $\sim$ into $MSPL$ resp. $ISPL$ in such a way that in the presence of **P**, **C**, and **M** we in fact obtain $\mathbf{N}^-$ resp. $\mathbf{N}$.[1] We shall point out a few well-known peculiarities of systems with constructive negation, notably the failure of intersubstitutivity of provable equivalents, and look at what after addition of $\sim$ happens to a number of previous concerns, viz. cut-elimination, decidability, interpolation, and the BHK interpretation of IPL. The BHK approach is extended from an interpretation of *one particular* system into a semantical framework for a broad *spectrum* of substructural logics.

6.1 Substructural subsystems of $\mathbf{N}^-$ and $\mathbf{N}$

Let $L^\sim$ denote the result of enriching the propositional language L by the new unary connective $\sim$ which is intended to denote *strong, constructive* negation. The notion of subformula for $L^\sim$ is defined in analogy to L, in particular every subformula of an $L^\sim$-formula A is also a subformula of $\sim A$.

Definition 6.1 (i) The rules of constructive minimal sequential propositional logic $COSPL^-$ (i.e. Nelson's constructive propositional logic $\mathbf{N}^-$ without structural inference rules) are the rules of $MSPL$ together with:

$$(\rightarrow\sim /)\quad X \rightarrow\sim B \quad Y \rightarrow A \vdash XY \rightarrow\sim (B/A);$$
$$(\sim / \rightarrow)\quad X \sim BAY \rightarrow C \vdash X \sim (B/A)Y \rightarrow C;$$

[1] Note that Fitch [1952] has developed formal systems with strong negation which are proper, though not substructural, subsystems of (the propositional part of) $\mathbf{N}^-$ resp. $\mathbf{N}$. They are obtained from the standard axiomatic presentation of $\mathbf{N}^-$ and $\mathbf{N}$ (cf. e.g. [Routley 1974]) by dropping the axiom-schema $(\sim (A \supset B) \supset (A \wedge \sim B)) \wedge ((A \wedge \sim B) \supset \sim (A \supset B))$, which syntactically captures the falsity conditions of implications $(A \supset B)$.

$(\rightarrow\sim\backslash)$ $X \rightarrow A \quad Y \rightarrow\sim B \vdash XY \rightarrow\sim (A \backslash B)$;
$(\sim\backslash\rightarrow)$ $XA \sim BY \rightarrow C \vdash X \sim (A \backslash B)Y \rightarrow C$;
$(\rightarrow\sim\circ)$ $X \rightarrow\sim A \quad Y \rightarrow\sim B \vdash XY \rightarrow\sim (A \circ B)$;
$(\sim\circ\rightarrow)$ $X \sim A \sim BY \rightarrow C \vdash X \sim (A \circ B)Y \rightarrow C$;
$(\rightarrow\sim\wedge)$ $X \rightarrow\sim A \vdash X \rightarrow\sim (A \wedge B)$,
$X \rightarrow\sim B \vdash X \rightarrow\sim (A \wedge B)$;
$(\sim\wedge\rightarrow)$ $X \sim AY \rightarrow C \quad X \sim BY \rightarrow C \vdash X \sim (A \wedge B)Y \rightarrow C$;
$(\rightarrow\sim\vee)$ $X \rightarrow\sim A \quad X \rightarrow\sim B \vdash X \rightarrow\sim (A \vee B)$;
$(\sim\vee\rightarrow)$ $X \sim BY \rightarrow C \vdash X \sim (A \vee B)Y \rightarrow C$,
$X \sim AY \rightarrow C \vdash X \sim (A \vee B)Y \rightarrow C$;
$(\rightarrow\sim\sim)$ $X \rightarrow A \vdash X \rightarrow\sim\sim A$;
$(\sim\sim\rightarrow)$ $XAY \rightarrow B \vdash X \sim\sim AY \rightarrow B$.

(ii) The rules of constructive sequential propositional logic $COSPL$ (i.e. Nelson's constructive propositional logic **N** without structural rules of inference) are those of $COSPL^-$ together with $(\bot\rightarrow)$ and:

$(\sim \mathbf{t} \rightarrow)$ $\vdash X \sim \mathbf{t}Y \rightarrow A$;
$(\sim \top \rightarrow)$ $\vdash X \sim \top Y \rightarrow A$;
$(\rightarrow\sim\bot)$ $\vdash X \rightarrow\sim \bot$.

The idea behind these sequent rules involving $\sim$ is that they directly reflect refutability conditions for main connectives or constants in the scope of $\sim$. If $(\bot\rightarrow)$, $(\sim \mathbf{t}\rightarrow)$, and $(\sim\top\rightarrow)$ are assumed, we say that $\bot$, $\sim\mathbf{t}$, and $\sim\top$ act or are treated as falsum constants; otherwise $\bot$, $\sim\mathbf{t}$, and $\sim\top$ are regarded as propositional variables. Adding all possible combinations of the earlier structural inference rules **P**, **C**, **C′**, **E**, **E′**, and **M** to $COSPL^-$ and $COSPL$, we obtain lattices of systems analogous to the families presented in Chapter 3. In $COSPL^-_\Delta$ resp. $COSPL_\Delta$ ($\Delta \subseteq \{\mathbf{P}, \mathbf{C}, \mathbf{C}', \mathbf{E}, \mathbf{E}', \mathbf{M}\}$), the intuitionistic minimal resp. intuitionistic negations $\neg^r$, $\neg^l$ are defined by $\neg^r A \stackrel{def}{=} (\bot/A)$, $\neg^l A \stackrel{def}{=} (A \backslash \bot)$.

Consider now the following translation τ from $L^\sim$ to the propositional language in $\supset$, $\wedge$, $\vee$, $\sim$, and $\bot$:

$$\begin{aligned}
\tau(q) &= q, && q \in PROP \cup \{\bot\}\\
\tau(\sim q) &= \sim q, && q \in PROP \cup \{\bot\}\\
\tau(\mathbf{t}) &= \tau(\top) = p \supset p, && \text{for some } p \in PROP\\
\tau(\sim \mathbf{t}) &= \tau(\sim\top) = \bot
\end{aligned}$$

$$\begin{aligned}
\tau(A \setminus B) &= \tau(B/A) = \tau(A) \supset \tau(B)\\
\tau(\sim (A \setminus B)) &= \sim \tau(A \setminus B)\\
\tau(\sim (B/A)) &= \sim \tau(B/A)\\
\tau(A \circ B) &= \tau(A \wedge B)\\
\tau(A \wedge B) &= \tau(A) \wedge \tau(B)\\
\tau(\sim (A \circ B)) &= \tau(\sim A) \wedge \tau(\sim B)\\
\tau(\sim (A \wedge B)) &= \sim \tau(A \wedge B)\\
\tau(A \vee B) &= \tau(A) \vee \tau(B)\\
\tau(\sim (A \vee B)) &= \sim \tau(A \vee B).
\end{aligned}$$

Note that τ is not compositional because of the clause for $\sim (A \circ B)$. The translation τ' maps formulas in $\supset$, $\wedge$, $\vee$, $\sim$, and $\bot$ to $L^{\sim}$-formulas; it distributes over the connectives which both languages have in common, and $\tau'(A \supset B) = \tau'(A) \setminus \tau'(B)$. The following observation justifies the identification $COSPL^{-}_{\{\mathbf{P,C,M}\}} = \mathbf{N}^{-}$:

Observation 6.2 (i) $\vdash_{COSPL^{-}_{\{\mathbf{P,C,M}\}}} A_1 \dots A_n \rightarrow A$ only if $\vdash_{\mathbf{N}^{-}} \tau(A_1) \dots \tau(A_n) \rightarrow \tau(A)$.
(ii) $\vdash_{\mathbf{N}^{-}} A_1 \dots A_n \rightarrow A$ only if $\vdash_{COSPL^{-}_{\{\mathbf{P,C,M}\}}} \tau'(A_1) \dots \tau'(A_n) \rightarrow \tau'(A)$.

PROOF By induction on the length of proofs. Here is one example concerning (i): $(\rightarrow\sim \circ)$. Suppose that $\vdash A_1 \dots A_n \rightarrow\sim A$, $\vdash B_1 \dots B_m \rightarrow\sim B$ and, by the induction hypothesis, $\vdash \tau(A_1) \dots \tau(A_n) \rightarrow \tau(\sim A)$, $\vdash \tau(B_1) \dots \tau(B_m) \rightarrow \tau(\sim B)$ in $\mathbf{N}^{-}$ resp. $\mathbf{N}$. Then, by $<\rightarrow \wedge>$, $\vdash \tau(A_1) \dots \tau(A_n)\tau(B_1) \dots \tau(B_m) \rightarrow \tau(\sim A) \wedge \tau(\sim B)$, i.e. $\vdash \tau(A_1) \dots \tau(A_n)\tau(B_1) \dots \tau(B_m) \rightarrow \tau(\sim (A \circ B))$. □

A well-known peculiarity of Nelson's systems $\mathbf{N}^{-}$ and $\mathbf{N}$ is the failure of intersubstitutivity of provable equivalents. Let again $(A \rightleftharpoons^{+} B)$ be defined as $(A \setminus B) \wedge (B \setminus A) \wedge (A/B) \wedge (B/A)$. Using the terminology of [Pearce & Rautenberg 1991], $\rightleftharpoons^{+}$ may be called acceptance-equivalence. In each system $COSPL^{-}_{\Delta}$ and $COSPL_{\Delta}$, provable acceptance-equivalence is an equivalence relation but *not* a congruence relation, i.e. one cannot prove the replacement theorem wrt it. E.g.: $\vdash_{COSPL^{-}} \rightarrow \sim (p \setminus q) \rightleftharpoons^{+} p \circ \sim q$, but $\nvdash_{COSPL^{-}} \rightarrow \sim\sim (p \setminus q) \rightleftharpoons^{+} \sim (p \circ \sim q)$ (see Chapter 9). Moreover, $\vdash_{COSPL} \rightarrow \bot \rightleftharpoons^{+} \sim A \wedge A$, but $\nvdash_{COSPL^{-}} \rightarrow\sim \bot \rightleftharpoons^{+} \sim (\sim A \wedge A)$. Rejection-equivalence $\rightleftharpoons^{-}$ can then be defined by $(A \rightleftharpoons^{-} B) \stackrel{def}{=} (\sim B \setminus \sim A) \wedge (\sim A \setminus \sim B) \wedge (\sim A / \sim B) \wedge (\sim B / \sim A)$. Analogously to $\vdash \rightarrow A \rightleftharpoons^{+} B$ iff $\vdash A \leftrightarrow B$, we have $\vdash \rightarrow A \rightleftharpoons^{-} B$ iff $\vdash \sim A \leftrightarrow \sim B$. Clearly, also provable rejection-equivalence fails to be a congruence relation in $COSPL^{-}_{\Delta}$ or $COSPL_{\Delta}$. If one defines strong equivalence $(A \rightleftharpoons B)$ as $(A \rightleftharpoons^{+} B) \wedge (A \rightleftharpoons^{-} B)$, then provable strong equivalence *is* a congruence relation in $COSPL^{-}_{\Delta}$ and $COSPL_{\Delta}$.[2] Let C_A denote an $L^{\sim}$-formula that contains a certain occurrence of A as a subformula, and let C_B denote the result of replacing this occurrence of A in C by B. The degree of

[2]The distinction between positive and negative (semantic) consequence is well-known from partial logic, see again e.g. [Fenstad, Halvorsen, Langholm & van Benthem 1987], [Thijsse 1990]. Note, however, that for the variety of notions of semantic consequence considered by Thijsse "logical equivalence [...] turns out as mutual consequence" as Thijsse [1990, p. 29] quotes from [Blamey 1986]. In other words, intersubstitutivity of provable equivalents holds.

A ($d(A)$) again is the number of occurrences of propositional constants and connectives (now including $\sim$) in A.

Theorem 6.3 (replacement) If $\to A \rightleftharpoons B$ is provable in $COSPL_{\Delta}^{-}$ or $COSPL_{\Delta}$, then so is $\to C_A \rightleftharpoons C_B$.

PROOF By induction on $l = d(C_A) - d(A)$. If $l = 0$, the proof is trivial. Assume that the claim holds for every $l \leq m$, and $l = m + 1$.

$C_A = \sim D$: Assume that $d(D_A) \leq l$ and $\vdash \to A \rightleftharpoons B$. By the induction hypothesis, $\vdash \to D_A \rightleftharpoons D_B$, and therefore the following formulas are provable: $D_A \setminus D_B$, $D_B \setminus D_A$, D_A/D_B, D_B/D_A, $\sim D_A\setminus \sim D_B$, $\sim D_B\setminus \sim D_A$, $\sim D_A/ \sim D_B$, and $\sim D_B/ \sim D_A$. By (*cut*), $(\setminus \to)$, $(\uparrow \setminus)$, $(/ \to)$, $(\uparrow /)$, $(\to\sim\sim)$, and $(\sim\sim\to)$, also $\sim\sim D_A\setminus \sim\sim D_B$, $\sim\sim D_B\setminus \sim\sim D_A$, $\sim\sim D_A/ \sim\sim D_B$, and $\sim\sim D_B/ \sim\sim D_A$ are provable and thus $\vdash \to C_A \rightleftharpoons C_B$.

$C_A = D_1 \nabla D_2$, $\nabla \in \{/, \setminus, \wedge, \circ, \vee\}$. We consider the case for $\nabla = \wedge$. Here we have the following derivations:

$$\dfrac{\dfrac{\dfrac{\dfrac{\to D_{1A} \setminus D_{1B}}{D_{1A} \to D_{1B}}}{D_{1A} \wedge D_2 \to D_{1B}} \qquad \dfrac{D_2 \to D_2}{D_{1A} \wedge D_2 \to D_2}}{D_{1A} \wedge D_2 \to D_{1B} \wedge D_2}}{\to C_A \setminus C_B;}$$

$$\dfrac{\dfrac{\dfrac{\dfrac{\to\sim D_{1A}\setminus \sim D_{1B}}{\sim D_{1A} \to\sim D_{1B}}}{\sim D_{1A} \to\sim (D_{1B} \wedge D_2)} \qquad \dfrac{\sim D_2 \to\sim D_2}{\sim D_2 \to\sim (D_{1B} \wedge D_2)}}{\sim (D_{1A} \wedge D_2) \to\sim (D_{1B} \wedge D_2)}}{\to\sim C_A\setminus \sim C_B.}$$

Analogously we obtain $C_B \setminus C_A$, $\sim C_B\setminus \sim C_A$, C_A/C_B, C_B/C_A, $\sim C_A/ \sim C_B$, and $\sim C_B/ \sim C_A$. The remaining cases are similar. □

The following collection of equivalences in terms of $\rightleftharpoons^+$ which are provable in $COSPL_{\Delta}^{-}$ without using (*cut*) will turn out useful:

$$(red1) \quad \sim (A \wedge B) \rightleftharpoons^+ (\sim A\vee \sim B), \quad \sim (A \vee B) \rightleftharpoons^+ (\sim A\wedge \sim B),$$

$$\sim (B/A) \rightleftharpoons^+ (\sim B \circ A), \qquad \sim (A \setminus B) \rightleftharpoons^+ (A\circ \sim B),$$

$$\sim (A \circ B) \rightleftharpoons^+ \sim A\circ \sim B, \qquad \sim\sim A \rightleftharpoons^+ A.$$

In $COSPL_{\Delta}$ also the following acceptance-equivalences are provable without resort to (*cut*):

$$(red2) \quad \sim \bot \rightleftharpoons^+ \mathbf{t},$$
$$\sim \mathbf{t} \rightleftharpoons^+ \bot,$$
$$\sim \top \rightleftharpoons^+ \bot.$$

These provable acceptance equivalences describe a procedure for associating to each $L^{\sim}$-formula A one $L^{\sim}$-formula B such that $\vdash_{COSPL_\Delta} \rightarrow A \rightleftharpoons^+ B$, resp. $\vdash_{COSPL_\Delta^-} \rightarrow A \rightleftharpoons^+ B$, and B has occurrences of $\sim$ only in front of propositional variables resp. propositional variables or constants, if for $COSPL_\Delta^-$, $\sim \bot$, $\sim \mathbf{t}$, and $\sim \top$, are associated to themselves.[3] Let us call this B the reduct of A, $r(A)$. Let $r'(A)$ denote the result of replacing in $r(A)$ each occurrence of $\sim q$ by q', for every $q \in PROP \cup \{\bot, \sim \mathbf{t}, \sim \top\}$. In $r'(A)$, $\sim$ does not occur. Consider now positive sequential propositional logic $PSPL$ (see Chapter 4) with $\{q' \mid q \in PROP \cup \{\bot, \mathbf{t}, \top\}\}$ as a set of fresh propositional variables. By induction on the length of proofs it can be shown that $COSPL_\Delta^-$ can be faithfully interpreted in $PSPL_\Delta$:[4]

Observation 6.4 $\vdash_{COSPL_\Delta^-} A_1 \ldots A_n \rightarrow A$ iff $\vdash_{PSPL_\Delta} r'(A_1) \ldots r'(A_n) \rightarrow r'(A)$.

This observation can reasonably be understood as expressing ***atomicity of strong negation*** in the basic systems $COSPL_\Delta^-$ of constructive logic: (i) *literals*, i.e. propositional variables and their strongly negated forms, can be identified as the basic building blocks of formulas, and (ii) if $\sim$ is 'pushed through' to the propositional variables and constants we arrive at (substructural subsystems of) positive propositional logic.

In $COSPL_\Delta^-$, $(A \wedge B) \rightleftharpoons^+ \sim (\sim A \vee \sim B)$ and $(A \vee B) \rightleftharpoons^+ \sim (\sim A \wedge \sim B)$ are provable (again without using (cut)). Together with $\vdash \;\sim (A \wedge B) \rightleftharpoons^+ \;\sim A \vee \sim B$, $\vdash \;\sim (A \vee B) \rightleftharpoons^+ \;\sim A \wedge \sim B$ this shows that in $COSPL_\Delta^-$ and $COSPL_\Delta$, $\wedge$ resp. $\vee$ can be defined by means of $\vee$ and $\sim$ resp. $\wedge$ and $\sim$. Moreover, from $(red2)$ we know that in $COSPL_\Delta$, $\bot$ can be defined as $\sim \mathbf{t}$.

The provable acceptance equivalences $(red1)$ and $(red2)$ specify the refutability (or rejectability) conditions referred to above. The refutability conditions for $(A \wedge B)$ resp. $(A \vee B)$ resp. $\sim A$ are identified as the provability conditions for $\sim A \vee \sim B$ resp. $\sim A \wedge \sim B$ resp. A, which is very natural. Also the refutability conditions for the directional implications are convincing, because in the absence of structural inference rules they are provability conditions of direction-sensitive, non-commutative $\circ$-conjunctions. Less clear are the rejectability conditions for $A \circ B$, since there is no 'intensional' disjunction corresponding to the 'intensional' conjunction $\circ$. So, what does it mean to refute a concatenation, i.e. a text, $A \circ B$? To assume that $\sim (A \circ B)$ is provably acceptance equivalent to $\sim A \vee \sim B$ is problematic, since we could then e.g. prove $A(A \setminus \sim A) \rightarrow \;\sim (A \circ (A \setminus \sim A))$ as well as $A(A \setminus \sim A) \rightarrow (A \circ (A \setminus \sim A))$. Moreover, $\vdash \;\rightarrow \;\sim (A \circ B) \rightleftharpoons^+ \;\sim A \wedge \sim B$ would, if no structural rules are assumed, make the provability conditions of a non-directional connective the refutability conditions of a direction-sensitive connective, which, as a kind of mismatch, would be rather surprising. In contrast to this, we may regard it as plausible that the refutation of a concatenation $A \circ B$ is provably acceptance equivalent to a refutation of each component of the concatenation: refuting a text means refuting every sentence in the text. The equivalences $(red1)$ may thus be viewed as a

[3] For the case of $\mathbf{N}$, this observation is due to Gurevich [1977]. Note that we cannot expect that to each $L^{\sim}$-formula A one can find a provably ***strongly equivalent***, and hence intersubstitutable, $L^{\sim}$-formula B such that in B, $\sim$ occurs only in front of propositional variables or constants. For instance, it is well-known that the implication $\supset$ is not definable in $\mathbf{N}^-$ or $\mathbf{N}$ by means of $\sim$, $\wedge$, $\vee$, and $\bot$ (cf. e.g. [Gurevich 1977]; his proof for $\mathbf{N}$ refers to [McKinsey 1939] and is also applicable to $\mathbf{N}^-$).

[4] For the case of $\mathbf{N}^-$ and positive propositional logic, this has been observed in [Pearce 1991].

justification of $COSPL_{\Delta}^{-}$'s sequent rules involving $\sim$; and indeed the very formulation of these rules is *induced* in an obvious way by ($red1$). The rejection equivalence of **t**, $\top$, and $\sim\bot$ in $COSPL_{\Delta}$ can be elaborated as follows: Since every sequence of premise occurrences proves **t**, **t** cannot be disproved; there is no sequence of premise occurrences that refutes **t**. Similarly, intuitionistic falsum cannot be proved; therefore in the constructive case, $\sim$ **t** should be added as a falsum. The verum constant $\top$ is a theorem. Although there are sequences X of premise occurrences such that $X \to \top$ is not provable in the absence of **M**, this does not mean that X refutes $\top$, i.e. $X \to\sim \top$ is provable. On the contrary, it is hardly imaginable that a theorem is refutable. In this way we arrive at the same refutability conditions for **t** and $\top$ and $\sim\bot$.

6.2 Cut-elimination, decidability, and interpolation

Before we come to cut-elimination as a key to decidability results, we point out that a well-known method for proving underivability of sequents in the $\{/, \backslash, \circ\}$-fragment of $MSPL$ and $MSPL_{\{\mathbf{P}\}}$, viz. p-count invariance (see [van Benthem 1986, 1991]), can straightforwardly be extended to the $\{/, \backslash, \circ, \sim\}$-fragment of $COSPL^{-}$ and $COSPL_{\{\mathbf{P}\}}^{-}$.

Definition 6.5 For every $p \in PROP$, the p-count of a formula A in $\{/, \backslash, \circ, \sim\}$ ($pc(A)$) is defined as follows:

$$
\begin{array}{lllll}
pc(p) & = & 1; & & \\
pc(q) & = & pc(\sim q) & = & 0, \qquad \text{if } p \neq q; \\
pc(A \backslash B) & = & pc(B/A) & = & pc(B) - pc(A); \\
pc(A \circ B) & = & pc(A) + pc(B); & & \\
pc(\sim p) & = & -pc(p); & & \\
pc(\sim (A \backslash B)) & = & pc(B/A) & = & pc(A) + pc(\sim B);
\end{array}
$$

$$
\begin{array}{lll}
pc(\sim (A \circ B)) & = & pc(\sim A) + pc(\sim B); \\
pc(\sim\sim A) & = & pc(A).
\end{array}
$$

For instance, $pc(\sim (p \backslash ((\sim p \backslash q) \backslash q))) = -2$.

Observation 6.6 If $\vdash_{COSPL^{-}_{/,\backslash,\circ,\sim}} A_1 \ldots A_n \to A$ or $\vdash_{COSPL^{-}_{/,\backslash,\circ,\sim\{\mathbf{P}\}}} A_1 \ldots A_n \to A$, then $pc(A_1) + \ldots + pc(A_n) = pc(A)$, for every $p \in PROP$.

PROOF By a straightforward induction on the length of proofs. Consider for instance ($\to\sim /$). If $\vdash A_1 \ldots A_n \to\sim B$, $\vdash B_1 \ldots B_m \to A$, then, by the induction hypothesis, $pc(A_1) + \ldots + pc(A_n) = pc(\sim B)$, $pc(B_1) + \ldots + pc(B_m) = pc(A)$. Hence, $pc(A_1) + \ldots + pc(A_n) + pc(B_1) + \ldots + pc(B_m) = pc(\sim B) + pc(A) = pc(\sim (B/A))$. □

The unprovability of $B/(A \backslash B) \to A$ e.g. shows that p-count invariance is only a necessary condition for provability of sequents in the $\{/, \backslash\}$-fragment of $MSPL$ and $MSPL_{\{\mathbf{P}\}}$ and hence also in the $\{/, \backslash, \circ, \sim\}$-fragment of $COSPL^{-}$ and $COSPL_{\{\mathbf{P}\}}^{-}$. Clearly, the structural rules **C**, **C'**, **E**, **E'**, and **M** spoil the p-count invariance property.

Let in what follows Ξ now also range over $COSPL^{-}$ and $COSPL$.

Theorem 6.7 (cut-elimination) Applications of (*cut*) can be eliminated from proofs in Ξ_Θ ($\Theta \subseteq \{\mathbf{P}, \mathbf{C}, \mathbf{C}', \mathbf{M}\}$).

PROOF Analogous to the proof of Theorem 3.4, now also taking into account occurrences of $\sim$. We shall by way of example just consider the rules for introducing $\sim\sim$ into premises and conclusions. The following list of conversion steps is exhaustive:

$$\left[\dfrac{\dfrac{\dfrac{\Pi_1}{XBY\to A}}{X{\sim\sim}BY\to A} \qquad \dfrac{\Pi_2}{X_1AX_2\to C}}{X_1X{\sim\sim}BYX_2\to C}\right] \text{ is converted into } \left[\dfrac{\dfrac{\dfrac{\Pi_1}{XBY\to A} \quad \dfrac{\Pi_2}{X_1AX_2\to C}}{X_1XBYX_2\to C}}{X_1X{\sim\sim}BYX_2\to C}\right]$$

$$\left[\dfrac{\dfrac{\Pi_1}{Y\vdash A} \qquad \dfrac{\dfrac{\Pi_2}{XAZ\to B}}{XAZ\to{\sim\sim}B}}{XYZ\to{\sim\sim}B}\right] \text{ is converted into } \left[\dfrac{\dfrac{\dfrac{\Pi_1}{Y\to A} \quad \dfrac{\Pi_2}{XAZ\to B}}{XYZ\to B}}{XYZ\to{\sim\sim}B}\right]$$

$$\left[\dfrac{\dfrac{\Pi_1}{Y\to A} \qquad \dfrac{\dfrac{\Pi_2}{X_1AX_2BZ\to C}}{X_1AX_2{\sim\sim}BZ\to C}}{X_1YX_2{\sim\sim}BZ\to C}\right] \text{ is converted into } \left[\dfrac{\dfrac{\dfrac{\Pi_1}{Y\to A} \quad \dfrac{\Pi_2}{X_1AX_2BZ\to C}}{X_1YX_2BZ\to C}}{X_1YX_2{\sim\sim}BZ\to C}\right]$$

$$\left[\dfrac{\dfrac{\dfrac{\Pi_1}{Y\to A}}{Y\to{\sim\sim}A} \qquad \dfrac{\dfrac{\Pi_2}{XAZ\to B}}{X{\sim\sim}AZ\to B}}{XYZ\to B}\right] \text{ is converted into } \left[\dfrac{\dfrac{\Pi_1}{Y\to A} \quad \dfrac{\Pi_2}{XAZ\to B}}{XYZ\to B}\right].\square$$

Corollary 6.8 (i) (subformula property) If $\vdash_{\Xi_\Theta} A_1 \ldots A_n \to A$, then there is a proof of this sequent in Ξ_Θ in which only subformulas of $A_1, \ldots, A_n$, and A occur;

(ii) If $\vdash_{\Xi_\Theta} \to A$, then there is a proof of this sequent in Ξ_Θ in which the last step is the application of a rule introducing a connective on the right side of $\to$;

(iii) (disjunction property) If $A \vee B$ is provable in Ξ_Θ, then A is provable or B is provable.

(iv) (constructible falsity) If $\sim (A \wedge B)$ is provable in $COSPL_\Theta^-$ or $COSPL_\Theta$, then $\sim A$ is provable or $\sim B$ is provable;

(v) Each subsystem obtained from Ξ_Θ by dropping all rules for certain constants or connectives is a conservative subsystem of Ξ_Θ;

(vi) $MSPL_\Theta$ is a conservative subsystem of $COSPL_\Theta^-$; $ISPL_\Theta$ is a conservative subsystem of $COSPL_\Theta$.

PROOF Obvious. □

Corollary 6.9 *Tertium non datur* in the forms $A \vee \neg^r A$, $A \vee \neg^l A$, and $A \vee \sim A$ is neither a theorem of $COSPL_\Delta^-$ nor of $COSPL_\Delta$.

Theorem 6.10 Ξ_Δ is not an n-valued logic, $1 \leq n < \omega$.

PROOF Exactly as the proof of Theorem 3.8. □.

Define the translation τ^*, like τ' above, except that

$$\tau^*(A \wedge B) \;=\; \begin{cases} \bot & \text{if } A \wedge B \text{ is of the form } C\wedge \sim C \text{ or } \sim C \wedge C \\ \tau^*(A) \wedge \tau^*(B) & \text{otherwise.} \end{cases}$$

Since the set of provable sequents is the same for **N** and **N** − (*cut*), the following observation justifies the identification $COSPL_{\{\mathbf{P},\mathbf{C},\mathbf{M}\}} = \mathbf{N}$:

Observation 6.11 (i) $\vdash_{COSPL_{\{\mathbf{P},\mathbf{C},\mathbf{M}\}}} A_1 \ldots A_n \rightarrow A$ only if $\vdash_{\mathbf{N}} \tau(A_1) \ldots \tau(A_n) \rightarrow \tau(A)$.
(ii) $\vdash_{\mathbf{N}-(cut)} A_1 \ldots A_n \rightarrow A$ only if $\vdash_{COSPL_{\{\mathbf{P},\mathbf{C},\mathbf{M}\}}} \tau^*(A_1) \ldots \tau^*(A_n) \rightarrow \tau^*(A)$.

PROOF By induction on the length of proofs. □

Theorem 6.12 Provability of sequents in Ξ_{Δ_1} is decidable ($\Delta_1 \subseteq \{\mathbf{P},\mathbf{M}\}$).

PROOF Exactly as the proof of Theorem 3.9. (Note that by "complexity" we could also still understand the number of occurrences of propositional constants and binary connectives, such that $\sim$ would not play a role here, and the rules ($\rightarrow\sim\sim$) and ($\sim\sim\rightarrow$) would not be considered as introducing complexity.) □

Theorem 6.13 Provability of sequents in Ξ_Θ is decidable.

PROOF The proof is the same as the proof of Theorem 3.11, except that in the construction of Ξ'_Θ for the systems with $\sim$ we also have the modified operational rule:

$$(\rightarrow\sim /)^0 \quad X \rightarrow\sim B \quad Y \rightarrow A \vdash [XY] \rightarrow\sim (B/A),$$

where $[XY] \rightarrow\sim (B/A)$ is the contraction of $XY \rightarrow\sim (B/A)$ such that any formula in XY occurs only 0 or 1 time fewer in $[XY]$ than in XY. In addition we have rules $(\rightarrow\sim \backslash)^0$, $(\sim / \rightarrow)^0$, $(\sim \backslash \rightarrow)^0$, $(\rightarrow\sim \circ)$, $(\sim \circ \rightarrow)^0$, $(\sim \wedge \rightarrow)^0$, and $(\sim \vee \rightarrow)^0$, which are likewise induced by the provable acceptance-equivalences ($red\,1$). □

We can now e.g. show that contraposition principles like $(p/q) \rightarrow (\sim q/ \sim p)$ fail to be provable in $COSPL_\Theta$ and $COSPL^-_\Theta$.

Let us eventually turn to interpolation. The notions of positive and negative occurrence of propositional variables in $L^\sim$-formulas, sequents built up from $L^\sim$-formulas and sequences of $L^\sim$-formula occurrences are defined as follows:

Definition 6.14 A propositional variable p occurs positively in the scope of an even number of occurrences of $\sim$; it occurs negatively in the scope of an uneven number of occurrences of $\sim$. A positive resp. negative occurrence of p in A remains positive resp. negative in $A \wedge B$, $B \wedge A$, $A \circ B$, $B \circ A$, $A \vee B$, $B \vee A$ and $X \rightarrow A$; the polarity of p in A is reversed in $\sim A$, $A \backslash B$, B/A, and $XAY \rightarrow B$. A propositional variable occurs positively resp. negatively in X iff it occurs positively resp. negatively in $\overset{\circ}{X}$. Let $pos(X)$ resp. $neg(X)$ denote the set of propositional variables that occur positively resp. negatively in X. A reversal of polarities in a sequence X is indicated by $\bar{X}$.

Theorem 6.15 (interpolation) If $\vdash_{\Xi_\Theta} X \rightarrow A$, then there is an $L^{\sim}$-formula C such that $\vdash_{\Xi_\Theta} X \rightarrow C$, $\vdash_{\Xi_\Theta} C \rightarrow A$, $pos(C) \subseteq (pos(X) \cap pos(A))$, and $neg(C) \subseteq (neg(X) \cap neg(A))$.

PROOF By the above observation that to each $L^{\sim}$-formula A one can find a provably acceptance-equivalent $L^{\sim}$-formula B with occurrences of $\sim$ only in front of propositional variables or constants such that A and B are interderivable, this interpolation theorem follows from Theorem 3.20. □

Corollary 6.16 (i) Interpolation holds for all elementary fragments of Ξ_Θ.
(ii) Interpolation in the sense of IND in Chapter 3 holds for the elementary fragments of Ξ and $\Xi_{\{M\}}$ based on: (a) $\{\sim, /, \circ\}$, $\{\sim, \backslash, \circ\}$, and $\{\sim, /, \backslash, \circ\}$, (b) every subset of $\{\sim, \circ, \wedge, \vee\}$, and the fragments obtained by joining any of the latter bases with one from (a).

6.3 The BHK interpretation reconsidered

The upshot of our discussion of the BHK interpretation of IPL in Chapter 2 was twofold: (a) Certain versions of the interpretation are somewhat ambiguous. One possible way of resolving this ambiguity is taking into account distinctions from the area of substructural logics, like the one between $\wedge$ and $\circ$. (b) In order to overcome problems resulting from the non-constructive nature of intuitionistic negation, the BHK interpretation in terms of proofs should be supplemented by an interpretation in terms of *disproofs*. We now take up this discussion again.

6.3.1 Ambiguity as providing degrees of freedom

In Chapter 2, we have emphasized certain ambiguities in various versions of the BHK interpretation of IPL. We may, however, also turn the tables and regard these ambiguities as *degrees of freedom*. This seems to be the proper methodological perspective in order to view the BHK interpretation as a *semantical framework* rather than an interpretation of one particular formal system. As often in the methodology of semantics a useful analogy emerges by considering Kripke's semantics for normal modal propositional logics.[5] The Kripke semantics forms a framework (or paradigm, or theory-ensemble) in the following sense: whereas the class of all Kripke frames characterizes the minimal normal modal propositional logic **K**, stronger calculi which may be appropriate for particular readings of the modal operators are characterized by imposing certain constraints on (the accessibility relation in) Kripke frames. The analogy for the case of the BHK interpretation would be this: there are BHK-like interpretations that constitute a semantics for certain basic logical systems; stronger calculi which may be appropriate for representing different conceptions of deductive information processing are then interpreted by imposing suitable conditions on the combination of proofs and disproofs.

In order to carry out this programme, we shall (i) lay down a proof/disproof interpretation for the logical constants and connectives $\perp$, **t**, $\top$, $\sim$, $/$, $\backslash$, $\wedge$, $\circ$, and $\vee$, (ii)

[5] See e.g. [Pearce & Wansing 1988].

show that this interpretation is sound wrt the basic constructive propositional logics $COSPL^-$ resp. $COSPL$ and (iii) specify additional conditions on the combination of proofs and disproofs, i.e. constraints on the juxtaposition operation, which correspond to the structural inference rules **P**, **C**, **C'**, and **M**. In order to prove soundness we need a semantical counterpart of the notion of provable sequent: the notion of ***strictly valid*** sequent.

Definition 6.17 A sequent $A_1 \dots A_n \rightarrow A$ is strictly valid iff the following holds:

$\forall \Pi_1 \dots \forall \Pi_n (pr(\Pi_1, A_1), \dots, pr(\Pi_n, A_n)$ implies $pr(\Pi_1 \dots \Pi_n, A))$, if $(1 \leq n)$;
$pr(<>, A)$, otherwise.

The notion of valid sequent, as used for the BHK interpretation of IPL, already ensures that each of the structural inference rules under consideration preserves validity. In the case of **M** e.g. it is enough to assume the existence of the operation of removing items from finite sequences. In contrast to this, the notion of strictly valid sequent by itself does not guarantee that the structural inference rules preserve strict validity. By the correspondence between a certain condition on the combination of proofs or disproofs and a certain structural rule R we mean that the condition ensures that R preserves strict validity.

6.3.2 The proof/disproof interpretation and its soundness wrt $COSPL_{\ominus}^-$ resp. $COSPL_{\ominus}$

In our proof/disproof interpretation we shall use $dpr(\Pi, A)$ to denote "Π is a disproof of A". We assume a universe of proofs and disproofs comprising the empty sequence $<>$. The proof/disproof interpretation consists of the following clauses:[6]

(1) $pr(<>, \top)$;

(2) $pr(\Pi_1 \dots \Pi_n, \mathbf{t})$ $(1 \leq n)$ iff $\exists A_1 \dots \exists A_n\ pr(\Pi_1, A), \dots, pr(\Pi_n, A_n)$;

(3) $pr(\Pi, \sim A)$ iff $dpr(\Pi, A)$;

(4) $pr(\Pi_1, (A/B))$ iff for every Π_2 such that $pr(\Pi_2, B)$, $pr(\Pi_1\Pi_2, A)$;

(5) $pr(\Pi_1, (B \setminus A))$ iff for every Π_2 such that $pr(\Pi_2, B)$, $pr(\Pi_2\Pi_1, A)$;

(6) $pr(\Pi, (A \circ B))$ iff $\exists \Pi_1 \exists \Pi_2$ $(\Pi_1\Pi_2 = \Pi$, $pr(\Pi_1, A)$, and $pr(\Pi_2, B))$;

(7) $pr(\Pi, (A \wedge B))$ iff $pr(\Pi, A)$ and $pr(\Pi, B)$;

(8) $pr(\Pi, (A \vee B))$ iff $pr(\Pi, A)$ or $pr(\Pi, B)$;

(9) $dpr(\Pi, \sim A)$ iff $pr(\Pi, A)$;

(10) $dpr(\Pi, (A/B))$ iff $\exists \Pi_1 \exists \Pi_2$ $(\Pi_1\Pi_2 = \Pi$, $dpr(\Pi_1, A)$, and $pr(\Pi_2, B))$;

(11) $dpr(\Pi, (B \setminus A))$ iff $\exists \Pi_1 \exists \Pi_2$ $(\Pi_1\Pi_2 = \Pi$, $pr(\Pi_1, B)$, and $dpr(\Pi_2, A))$;

[6] The constant $\perp$ and $\sim \mathbf{t}$, $\sim \top$ are treated as propositional variables.

(12) $dpr(\Pi, (A \circ B))$ iff $\exists\Pi_1\exists\Pi_2\,(\Pi_1\Pi_2 = \Pi, dpr(\Pi_1, A)$, and $dpr(\Pi_2 B))$;

(13) $dpr(\Pi, (A \wedge B))$ iff $dpr(\Pi, A)$ or $dpr(\Pi, B)$;

(14) $dpr(\Pi, (A \vee B))$ iff $dpr(\Pi, A)$ and $dpr(\Pi, B)$.

Theorem 6.18 $COSPL^-$ is sound wrt the proof/disproof interpretation, i.e. every sequent provable in $COSPL^-$ is strictly valid.

PROOF By a straightforward induction on the length of proofs in $COSPL^-$. The logical rules (*id*) and (*cut*) for which there are no interpreting clauses are covered by the definition of strict validity of sequents. □

It can readily be verified that the following conditions on the combination of proofs or disproofs correspond to the structural inference rules they are associated with:

P $pr(\Pi_1\Pi_2\Pi_3\Pi_4, A)$ implies $pr(\Pi_1\Pi_3\Pi_2\Pi_4, A)$;

C $pr(\Pi_1\Pi_2\Pi_3\Pi_4, A), pr(\Pi_2, B), pr(\Pi_3, B)$ implies $pr(\Pi_1\Pi_2\Pi_4, A)$, and $pr(\Pi_1\Pi_3\Pi_4, A)$;

C′ $pr(\Pi_1\Pi_2\Pi_3\Pi_4\Pi_5, A), pr(\Pi_2, B), pr(\Pi_4, B)$ implies $pr(\Pi_1\Pi_3\Pi_4\Pi_5, A)$, and $pr(\Pi_1\Pi_2\Pi_3\Pi_5, A)$;

M $pr(\Pi_1\Pi_3, B), pr(\Pi_2, A)$ implies $pr(\Pi_1\Pi_2\Pi_3, B)$.

Theorem 6.19 $COSPL_\Theta^-$ is sound wrt the proof/disproof interpretation extended by the conditions that correspond to the structural rules in Θ, i.e. every sequent provable in $COSPL_\Theta^-$ is strictly valid.

PROOF Again by a straightforward induction. □

In $COSPL_\Theta$, $\bot$, $\sim \mathbf{t}$, and $\sim \top$ act as intuitionistic falsum constants. In this case we have the following additional clauses:

(15) $pr(\Pi, \bot)$ iff $dpr(\Pi, \mathbf{t})$ iff $dpr(\Pi, \top)$;

(16) $dpr(\Pi, \bot)$ iff $pr(\Pi, \mathbf{t})$.

The semantical counterpart of $(\bot \rightarrow)$, $(\sim \mathbf{t} \rightarrow)$ and $(\sim \top \rightarrow)$ is the assumption that there is no construction which proves $\bot$, i.e. using this assumption, $(\bot \rightarrow)$, $(\sim \mathbf{t} \rightarrow)$ and $(\sim \top \rightarrow)$ preserve strict validity.

Introducing the BHK interpretation, by a proof we agreed to understand a canonical (or direct), (*cut*)-free proof. Therefore the above considerations have been restricted to $COSPL_\Theta^-$ resp. $COSPL_\Theta$, which we know to admit of cut-elimination.

Although the BHK approach towards IPL is usually referred to as an ***interpretation***, the less formal term 'BHK explanation' is widely felt to be the more appropriate terminology (cf. e.g. [Sundholm 1983]). Probably for this reason the question whether IPL is ***complete*** wrt the BHK interpretation has hardly ever been raised. The proof/disproof format developed above, however, seems explicit enough for seriously asking questions like: "Is $COSPL_\Theta^-$ complete wrt the proof/disproof interpretation extended by the conditions that correspond to the structural rules in Θ?".

[illegible]

Theorem 8.[illegible] [illegible] proof/[illegible] interpretation [illegible]

[illegible] for which there [illegible] the [illegible]

[illegible] the formalization of proofs [illegible] with [illegible]

[illegible]

Theorem 8.15 [illegible] the [illegible] extension [illegible] proves [illegible] is strictly weaker [illegible]

[illegible]

[illegible] In this case we [illegible] the following [illegible]

[illegible]

Although the [illegible] terms [illegible] proof [illegible] in the [illegible]

Chapter 7

Functional completeness for substructural subsystems of N

Considering in addition to the notion of *proof* the corresponding notion of *refutation* (or disproof), von Kutschera [1969] has extended his earlier functional completeness result for IPL and has shown that the connectives $\wedge$, $\vee$, $\supset$, and $\neg$ of what he calls *direct propositional logic* and *extended direct propositional logic* are functionally complete wrt to generalizations of his earlier [1968] proof-theoretic semantics. As can immediately be verified, direct resp. extended direct propositional logic is exactly Nelson's constructive propositional system $\mathbf{N}^-$ resp. $\mathbf{N}$ (cf. Chapter 2 or [Almukdad & Nelson 1984]), if von Kutschera's $\neg$ is translated as $\sim$. In this chapter, we will show how von Kutschera's proof of functional completeness for $\mathbf{N}^-$ and $\mathbf{N}$ can be extended to $COSPL_\Delta^-$ and $COSPL_\Delta$. The functional completeness results for $COSPL_\Delta^-$ and $COSPL_\Delta$ together with the functional completeness results of Chapter 4 will become integral for arguing to the effect that the structures used in the monoid semantics of Chapter 9 represent in a certain sense an *exhaustive* format of abstract information structures.

7.1 Disproofs as mirror-images of proofs

Von Kutschera's approach to the problem of functional completeness for direct propositional logic, i.e. (the propositional part of) $\mathbf{N}^-$, is based on the following considerations (formulated wrt an axiomatic setting):

> Die Einführung des Widerlegungsbegriffs für die Gentzenkalküle, nach der eine Formel widerlegbar ist, wenn aus ihr beliebige Formeln ableitbar sind, ist ... keineswegs zwingend. Ebenso hätte man z.B. von einem Widerlegungsbegriff ausgehen können, der in Analogie zum ... Beweisbegriff eingeführt ist, d.h. man hätte von Kalkülen K ausgehen können, die durch Antiaxiome definiert sind, die in K widerlegbar sind, und durch Deduktionsregeln, die besagen, wie aus bereits in K widerlegten Formeln eine neue in K widerlegbare Formel gewonnen werden kann. Dann hätte man den Beweisbegriff so einführen können, daß eine Formel beweisbar ist, wenn durch eine Widerlegung dieser Formel beliebige Formeln widerlegt werden können. So würde man zu einer Logik gelangen, die sich zur intuitionistischen

Logik gewißermaßen spiegelbildlich bezüglich Beweis- und Widerlegungsbegriff verhält. [Kutschera 1969, p. 104].[1]

The idea is thus to consider provability and refutability as a pair of prima facie independent and equally important primitive notions. As far as refutability is concerned, this is an early version of Słupecki's notion of *inverse consequence*; see [Słupecki, Bryll & Wybraniec-Skardowska 1972, 1973]. Instead of directly considering in addition to the notion of proof the notion of refutation, von Kutschera proceeds as follows: (i) every antiaxiom A of K is translated into the formula $-A$,[2] (ii) inverse inference rules $A_1, \ldots, A_n \vdash A$ which proceed from the refutability of $A_1, \ldots, A_n$ to the refutability of A are recast as $-A_1, \ldots, -A_n \vdash -A$, (iii) if A is refutable in K, then $-A$ is considered to be provable in K, and (iv) $--A$ is replaced by A. In this way, the ordinary notion of an axiomatic calculus may be used. Now, to an axiomatic calculus K one can associate a sequent calculus S_K by translating each K-axiom A into the S_K-rule $\vdash \ \rightarrow A$ and translating each K-rule $X \vdash A$ into the S_K-rule $\vdash X \rightarrow A$. Thus, working with sequent calculi, one obtains a language extended by a 'structural' negation operation – in the sense of refutation or disproof.

7.2 The higher-level Gentzen calculus $\mathbf{GN}^-$

In this section we shall introduce the higher-level sequent calculus $\mathbf{GN}^-$ as the underlying proof-theoretic framework for introducing connectives into premises and conclusions. Let again L be any formal language and let $FORM(\mathsf{L})$ be the set of all L-formulas. In order to be able to extend von Kutschera's approach, we have to make stronger assumptions and need an underlying sequent calculus which in addition to $\leftarrow$, $\rightarrow$ not only makes use of $-$ but also of a 'structural' connective $\odot$ corresponding to $\circ$.

Definition 7.1 The set of all RL-formulas is the smallest set Γ such that

$FORM(\mathsf{L}) \subseteq \Gamma$;

if $T, U \in \Gamma$, then $(T \odot U) \in \Gamma$;

if $T \in \Gamma$, then $(\rightarrow T), (T \leftarrow) \in \Gamma$;

if $T_1, \ldots T_n, U \in \Gamma$, then $(T_1 \ldots T_n \rightarrow U), (U \leftarrow T_1 \ldots T_n) \in \Gamma$;

[1] "Introducing the notion of refutation for the Gentzen calculi by saying that a formula is refutable, if arbitrary formulas can be inferred from it, is ... by no means obligatory. One might as well have started with e.g. a notion of refutation introduced in analogy to the ... notion of proof, i.e. with calculi K which are defined by antiaxioms refutable in K and by inference rules specifying how new formulas refutable in K can be obtained from formulas which have already been refuted in K. Then the notion of proof might have been introduced in such a way that a formula is provable, if by a refutation of this formula arbitrary formulas become refutable. This way one would end up with a logic which is, so to speak, a mirror-image of intutitionistic logic wrt the notions of proof and refutation." (translation HW)

[2] Actually, instead of '$-$' von Kutschera uses '$\sim$', which we use to denote strong negation in the object language, whereas von Kutschera denotes strong negation by '$\neg$'.

if $T \in \Gamma$, then $-T \in \Gamma$.[3]

We shall use T, U, T_1, T_2, ... resp. X, Y, X_1, X_2, ... to denote RL-formulas resp. finite, possibly empty sequences of RL-formula occurences.

Definition 7.2 Every $A \in FORM(\mathsf{L})$ is an RL-formula of R-degree 0;

if T has R-degree n, then $-T$ has R-degree n;

if n is the maximum of the R-degrees of the RL-formulas T, U, then the R-degree of $(T \odot U)$ is $n+1$;

if n is the maximum of the R-degrees of the RL-formulas in X and the R-degree of T, then the R-degree of $\mathsf{X} \rightarrow T$, $T \leftarrow \mathsf{X}$ is $n+1$.

If the R-degree of $T = n$, we write $Rd(T) = n$. An RL-formula U of the form $\mathsf{X} \rightarrow T$ or $T \leftarrow \mathsf{X}$ is said to be a higher-level sequent, if U contains more than one occurrence of $\rightarrow$ or $\leftarrow$, otherwise U is called a sequent.

Definition 7.3 Every RL-formula is an R-subformula of itself;

every R-subformula of T is an R-subformula of $-T$;

every R-subformula of T and every R-subformula of U is an R-subformula of $(T \odot U)$;

every R-subformula of T and every R-subfomula in X is an R-subformula of $\mathsf{X} \rightarrow T$, $T \leftarrow \mathsf{X}$.

The R-subformulas of T of R-degree 0 are called formula components of T. Let $T_1 \ldots T_n \Rightarrow T$ resp. $\Rightarrow T$ abbreviate $T_1 \ldots T_n \rightarrow T$ and $T \leftarrow T_1 \ldots T_n$ resp. $\rightarrow T$ and $T \leftarrow$.

The calculus $\mathbf{GN}^-$ is an extension of the result of removing $(\bot \Rightarrow)$ from the calculus **G** in Chapter 4. What is new are the rules for $\odot$ and the specification of refutability conditions for possibly higher-level sequents. It seems natural to say that $\mathsf{X} \Rightarrow T$ resp. $\mathsf{X} \Rightarrow -T$ is refutable iff on the strength of the provability of each occurrence in X, T is refutable resp. provable. We then obtain:

Definition 7.4 The rules of $\mathbf{GN}^-$ are those of the earlier higher-level Gentzen calculus **G** apart from $(\bot \Rightarrow)$, together with:

[3] Apart from using $\odot$, this definition deviates from von Kutschera's definition of R-formulas [1969, p. 105] also insofar as von Kutschera doesn't allow iterations of the structural negation. See, however, his remark on p. 106 "setzen wir fest, daß $\sim S$ für U stehen soll, wo S mit $\sim U$ identisch ist" ("we stipulate that $\sim S$ stands for U, where S is identical with $\sim U$" (translation HW)).

$$
\begin{array}{ll}
(\rightarrow - \leftarrow) & (\mathsf{X} \rightarrow -T)(\mathsf{Y} \rightarrow U) \vdash \mathsf{XY} \rightarrow -(T \leftarrow U);\\
(- \leftarrow\rightarrow) & \mathsf{X} - TU\mathsf{Y} \rightarrow T_1 \vdash \mathsf{X} - (T \leftarrow U)\mathsf{Y} \rightarrow T_1;\\
(\rightarrow - \rightarrow) & (\mathsf{X} \rightarrow U)(\mathsf{Y} \rightarrow -T) \vdash \mathsf{XY} \rightarrow -(U \rightarrow T);\\
(- \rightarrow\rightarrow) & \mathsf{X}U - T\mathsf{Y} \rightarrow T_1 \vdash \mathsf{X} - (U \rightarrow T)\mathsf{Y} \rightarrow T_1;\\
(- \leftarrow\leftarrow) & (-T \leftarrow \mathsf{X})(U \leftarrow \mathsf{Y}) \vdash -(T \leftarrow U) \leftarrow \mathsf{XY};\\
(\leftarrow - \leftarrow) & T_1 \leftarrow \mathsf{X} - TU\mathsf{Y} \vdash T_1 \leftarrow \mathsf{X} - (T \leftarrow U)\mathsf{Y};\\
(- \rightarrow\leftarrow) & (U \leftarrow \mathsf{X})(-T \leftarrow \mathsf{Y}) \vdash -(U \rightarrow T) \leftarrow \mathsf{XY};\\
(\leftarrow - \rightarrow) & T_1 \leftarrow \mathsf{X}U - T\mathsf{Y} \vdash T_1 \leftarrow \mathsf{X} - (U \rightarrow T)\mathsf{Y};\\
(\rightarrow \odot) & (\mathsf{X} \rightarrow T_1)(\mathsf{Y} \rightarrow T_2) \vdash \mathsf{XY} \rightarrow (T_1 \odot T_2);\\
(\odot \leftarrow) & (T_1 \leftarrow \mathsf{X})(T_2 \leftarrow \mathsf{Y}) \vdash (T_1 \odot T_2) \leftarrow \mathsf{XY};\\
(\odot \rightarrow) & \mathsf{X}TU\mathsf{Y} \rightarrow T_1 \vdash \mathsf{X}(T \odot U)\mathsf{Y} \rightarrow T_1;\\
(\leftarrow \odot) & T_1 \leftarrow \mathsf{X}TU\mathsf{Y} \vdash T_1 \leftarrow \mathsf{X}(T \odot U)\mathsf{Y};\\
(\rightarrow -\odot) & (\mathsf{X} \rightarrow -T_1)(\mathsf{Y} \rightarrow -T_2) \vdash \mathsf{XY} \rightarrow -(T_1 \odot T_2);\\
(-\odot \leftarrow) & (-T_1 \leftarrow \mathsf{X})(-T_2 \leftarrow \mathsf{Y}) \vdash -(T_1 \odot T_2) \leftarrow \mathsf{XY};\\
(-\odot \rightarrow) & \mathsf{X} - T - U\mathsf{Y} \rightarrow T_1 \vdash \mathsf{X} - (T \odot U)\mathsf{Y} \rightarrow T_1;\\
(\leftarrow -\odot) & T_1 \leftarrow \mathsf{X} - T - U\mathsf{Y} \vdash T_1 \leftarrow \mathsf{X} - (T \odot U)\mathsf{Y}.
\end{array}
$$

Note that (i) the rules $(\rightarrow - \leftarrow)$ - $(\leftarrow - \rightarrow)$ parallel the rules $(\rightarrow\!\sim \backslash)$, $(\rightarrow\!\sim /)$, $(\sim \backslash \rightarrow)$, and $(\sim / \rightarrow)$, and (ii) the rules involving $\odot$ parallel the earlier sequent rules for $\circ$. It can readily be verified that $(\odot \rightarrow)$ and $(\leftarrow \odot)$ are equivalent to, i.e. interreplaceable with,

$$
\begin{array}{ll}
(\odot \Uparrow) & \mathsf{X}(T_1 \odot T_2)\mathsf{Y} \rightarrow U \vdash \mathsf{X}T_1T_2\mathsf{Y} \rightarrow U,\\
& U \leftarrow \mathsf{X}(T_1 \odot T_2)\mathsf{Y} \vdash U \leftarrow \mathsf{X}T_1T_2\mathsf{Y}.
\end{array}
$$

$\mathcal{D}_{\mathbf{GN}^-}(\Pi, T, \mathsf{X})$, "$\Pi$ is a derivation in $\mathbf{GN}^-$ of T from the finite, possibly empty sequence X of $R\mathsf{L}$-formula occurrences", is defined in analogy to the notion of derivation in the higher-level sequent calculus **G** in Chapter 4.

Let U_V be an $R\mathsf{L}$-formula that contains a certain occurence of V as an R-subformula, and let U_T be the result of replacing this occurrence of V in U by T. Let $V \Leftrightarrow T$ denote $V \Rightarrow T$ and $T \Rightarrow V$, and let $V \Leftrightarrow^s T$ ("V and T are strongly interderivable") denote $V \Leftrightarrow T$, and $-T \Leftrightarrow -V$.

Theorem 7.5 If $\vdash V \Leftrightarrow^s T$ in $\mathbf{GN}^-$, then $\vdash U_V \Leftrightarrow^s U_T$ in $\mathbf{GN}^-$.

Proof By induction on $n = Rd(U_V) - Rd(V)$, analogous to the proof of Theorem 4.5. Here we consider the case that $U_V = (T_{1V} \odot T_2)$ or $(T_1 \odot T_{2V})$. Assume that the claim holds for every $n \leq m$, $n = m + 1$, T_{1V} resp. T_{2V} contains the occurrence of V in question, and $Rd(T_{1V}) \leq n$ resp. $Rd(T_{2V}) \leq n$. Now suppose that $\vdash V \Leftrightarrow^s T$. By applying $(\odot \rightarrow)$, $(\leftarrow \odot)$, $(\odot \Uparrow)$, and (tra) to $U_V \Leftrightarrow^s U_V$ we obtain $\vdash U_V \Leftrightarrow^s U_T$. □

7.3 Generalized Gentzen semantics

The idea is to characterize an n-ary connective F in the language L of a given logic $\mathcal{L}$ by the inference rules used to introduce $F(A_1, \dots, A_n)$ and now also $-F(A_1, \dots, A_n)$ into premises and conclusions. The proof-theoretic framework for specifying such inference rules will be **GN**$^-$ and the semantics is again codified in certain general rule schemata for formulating the inference rules. These schemata are subject to our earlier constraints, which now read as follows:

(i) Rule schemata characterizing F mention apart from one occurrence of F no other occurrence of a propositional connective; the role of formulas $F(A_1, \dots, A_n)$ and $-F(A_1, \dots, A_n)$ in deductive contexts depends on the deductive relationships between $A_1, \dots, A_n$ only.

(ii) The rule schemata for F are non-creative ones, i.e. every proof of an F-free formula A in the result of extending **GN**$^-$ by instantiations of these schemata can be converted into a proof of A with no applications of rules characterizing F.

As in Chapter 4, we obtain the following rule schemata for introducing $F(A_1, \dots, A_n)$ into premises and conclusions:

$$
\begin{array}{ll}
(I)\,(a) & \mathsf{X}_{11}\mathsf{W}_{11}\mathsf{Y}_{11} \rightarrow T_{11} \;\dots\; \mathsf{X}_{1s_1}\mathsf{W}_{1s_1}\mathsf{Y}_{1s_1} \rightarrow T_{1s_1} \vdash \mathsf{W}_{11}\dots\mathsf{W}_{1s_1} \rightarrow F(A_1, \dots, A_n), \\
& \vdots \\
& \mathsf{X}_{t1}\mathsf{W}_{t1}\mathsf{Y}_{t1} \rightarrow T_{t1} \;\dots\; \mathsf{X}_{ts_t}\mathsf{W}_{ts_t}\mathsf{Y}_{ts_t} \rightarrow T_{ts_t} \vdash \mathsf{W}_{t1}\dots\mathsf{W}_{ts_t} \rightarrow F(A_1, \dots, A_n), \\
& T_{11} \leftarrow \mathsf{X}_{11}\mathsf{W}_{11}\mathsf{Y}_{11} \;\dots\; T_{1s_1} \leftarrow \mathsf{X}_{1s_1}\mathsf{W}_{1s_1}\mathsf{Y}_{1s_1} \vdash F(A_1, \dots, A_n) \leftarrow \mathsf{W}_{11}\dots\mathsf{W}_{1s_1}, \\
& \vdots \\
& T_{t1} \leftarrow \mathsf{X}_{t1}\mathsf{W}_{t1}\mathsf{Y}_{t1} \;\dots\; T_{ts_t} \leftarrow \mathsf{X}_{ts_t}\mathsf{W}_{ts_t}\mathsf{Y}_{ts_t} \vdash F(A_1, \dots, A_n) \leftarrow \mathsf{W}_{t1}\dots\mathsf{W}_{ts_t}; \\
(I)\,(b) & \mathsf{X}_1\mathsf{W}\mathsf{Y}_1 \rightarrow T_1 \dots \mathsf{X}_j\mathsf{W}\mathsf{Y}_j \rightarrow T_j \vdash \mathsf{W} \rightarrow F(A_1, \dots, A_n), \\
& T_1 \leftarrow \mathsf{X}_1\mathsf{W}\mathsf{Y}_1 \dots T_j \leftarrow \mathsf{X}_j\mathsf{W}\mathsf{Y}_j \vdash F(A_1, \dots, A_n) \leftarrow \mathsf{W}; \\
& \\
(II)\,(a) & \mathsf{Y}_1\mathsf{X}_1\mathsf{Y}_2 \rightarrow T \;\dots\; \mathsf{Y}_1\mathsf{X}_t\mathsf{Y}_2 \rightarrow T \vdash \mathsf{Y}_1 F(A_1, \dots, A_n)\mathsf{Y}_2 \rightarrow T, \\
& T \leftarrow \mathsf{Y}_1\mathsf{X}_1\mathsf{Y}_2 \;\dots\; T \leftarrow \mathsf{Y}_1\mathsf{X}_t\mathsf{Y}_2 \vdash T \leftarrow \mathsf{Y}_1 F(A_1, \dots, A_n)\mathsf{Y}_2; \\
(II)\,(b) & \mathsf{Y}_1(\mathsf{X}_l \rightarrow (T_l \leftarrow \mathsf{Y}_l))\mathsf{Y}_2 \rightarrow U \vdash \mathsf{Y}_1 F(A_1, \dots, A_n)\mathsf{Y}_2 \rightarrow U, \\
& U \leftarrow \mathsf{Y}_1(\mathsf{X}_l \rightarrow (T_l \leftarrow \mathsf{Y}_l))\mathsf{Y}_2 \vdash U \leftarrow \mathsf{Y}_1 F(A_1, \dots, A_n)\mathsf{Y}_2, \quad l = 1, \dots, j.
\end{array}
$$

The W_{ik_i} $(i = 1, \dots, t; k_i = 1, \dots, s_i)$, $\mathsf{Y}_1, \mathsf{Y}_2$ and W are unspecified sequences of RL-formula occurrences, whereas $\mathsf{X}_{ik_i}, \mathsf{Y}_{ik_i}, T_{ik_i}$ resp. X_1, …, X_j, Y_1, …, Y_j, T_1, …, T_j contain only formula components from $A_1, \dots, A_n$. Moreover, in each instantiation of (I) (a) resp. (I) (b) every A_k $(k = 1, \dots, n)$ occurs in some X_{ik_i}, Y_{ik_i}, or T_{ik_i} resp. in some X_l, Y_l, or T_l $(l = 1, \dots, j)$. $\mathsf{X}_i = \mathsf{X}_{i1} \rightarrow (T_{i1} \leftarrow \mathsf{Y}_{i1}) \dots \mathsf{X}_{is_i} \rightarrow (T_{is_i} \leftarrow \mathsf{Y}_{is_i})$. If $n = 0$, then (I) (a) is $\vdash \Rightarrow F$, (I) (b) is $\vdash \mathsf{W} \Rightarrow F$, (II) (a) is $\mathsf{Y}_1\mathsf{Y}_2 \rightarrow U \vdash \mathsf{Y}_1 F \mathsf{Y}_2 \rightarrow U$, $U \leftarrow \mathsf{Y}_1\mathsf{Y}_2 \vdash U \leftarrow \mathsf{Y}_1 F \mathsf{Y}_2$, and (II) (b) is not instantiated.

In Chapter 4 we have noted that the schemata (I) (a) resp. (I) (b) are equivalent to the schemata

$(I)'(a)$ $\mathsf{W}_{i1} \rightarrow (\mathsf{X}_{i1} \rightarrow (T_{i1} \leftarrow \mathsf{Y}_{i1})) \ldots \mathsf{W}_{is_i} \rightarrow (\mathsf{X}_{is_i} \rightarrow (T_{is_i} \leftarrow \mathsf{Y}_{is_i})) \vdash$
$\vdash \mathsf{W}_{i1} \ldots \mathsf{W}_{is_i} \rightarrow F(A_1, \ldots, A_n),$
$(\mathsf{X}_{i1} \rightarrow (T_{i1} \leftarrow \mathsf{Y}_{i1})) \leftarrow \mathsf{W}_{i1} \ldots (\mathsf{X}_{is_i} \rightarrow (T_{is_i} \leftarrow \mathsf{Y}_{is_i})) \leftarrow \mathsf{W}_{is_i} \vdash$
$\vdash F(A_1, \ldots, A_n) \leftarrow \mathsf{W}_{i1} \ldots \mathsf{W}_{is_i}, \quad i = 1, \ldots, t;$

$(I)'(b)$ $\mathsf{W} \rightarrow (\mathsf{X}_1 \rightarrow (T_1 \leftarrow \mathsf{Y}_1)) \ldots \mathsf{W} \rightarrow (\mathsf{X}_j \rightarrow (T_j \leftarrow \mathsf{Y}_j)) \vdash$
$\vdash \mathsf{W} \rightarrow F(A_1, \ldots, A_n),$
$(\mathsf{X}_1 \rightarrow (T_1 \leftarrow \mathsf{Y}_1)) \leftarrow \mathsf{W} \ldots (\mathsf{X}_j \rightarrow (T_j \leftarrow \mathsf{Y}_j)) \leftarrow \mathsf{W} \vdash$
$\vdash F(A_1, \ldots, A_n) \leftarrow \mathsf{W}.$

As far as rule schemata for introducing $-F(A_1, \ldots, A_n)$ into premises and conclusions are concerned, we shall for two reasons significantly deviate from von Kutschera's [1969] approach. (i) As it seems, the schemata for introductions into conclusions as eventually presented by von Kutschera fail to be the exact realization of the result of his motivating considerations. Von Kutschera states that

> Es liegt ... nahe, die Regelschemata zur Einführung von $\sim F(A_1, \ldots, A_n)$ als Hinterformel so zu wählen, daß, wenn nach (I) aus der Beweisbarkeit von $S_{i1}, \ldots, S_{is_i}$ für ein i die Beweisbarkeit von $F(A_1, \ldots, A_n)$ folgt, nun aus der Widerlegbarkeit aller S-Formelreihen [what is meant is "R-Formelreihen", HW] $S_{i1}, \ldots, S_{is_i}$ die Widerlegbarkeit von $F(A_1, \ldots, A_n)$ folgt. [1969, p. 108][4]

Given this, admittedly not completely unequivocal, explanation, von Kutschera's [1969, p. 108] schemata are quite surprising:

$$\Delta \rightarrow\sim S_{1k_{1l}}; \ldots; \Delta \rightarrow\sim S_{tk_{t1}} \vdash \Delta \rightarrow\sim F(A_1, \ldots, A_n),$$
$$\ldots$$
$$\Delta \rightarrow\sim S_{1k_{1r}}; \ldots; \Delta \rightarrow\sim S_{tk_{tr}} \vdash \Delta \rightarrow\sim F(A_1, \ldots, A_n),$$

where Δ is a set of unspecified R-formulas and $r = s_1 \times \ldots \times s_t$ ($\times$ denoting multiplication) and $k_{il} = 1, \ldots, s_i$ for $l = 1, \ldots, r$. These schemata capture the rules for the connectives considered by von Kutschera; but note that in the absence of the standard structural inference rules von Kutschera need not distinguish between $\wedge$ and $\circ$.

(ii) Von Kutschera's schema for introducing $\sim F(A_1, \ldots, A_n)$ into premises turns out to be appropriate for the proof of functional completeness only because monotonicity of inference holds for his generalized Gentzen calculus. (See [1969, p. 110], where it says that finally we have:

"$\sim S_{1k_{1l}}, \ldots, \sim S_{tk_{tl}} \rightarrow\sim S_{k_{1l}}$ [what is meant is "$\sim S_{1k_{1l}}$", HW]; ...".)

We shall adopt von Kutschera's motivating ideas, but render them (more) *literally* into introduction schemata. We agree that one would like to avoid that both

[4] It is obvious ... that the rule schemata for introducing $\sim F(A_1, \ldots, A_n)$ into conclusions has to be chosen in such a way that, if the provability of $F(A_1, \ldots, A_n)$ follows from the provability of $S_{i1}, \ldots, S_{is_i}$ for some i, then the refutability of $F(A_1, \ldots, A_n)$ follows from the refutability of every succession of S-formulas [what is meant is "R-formulas", HW] $S_{i1}, \ldots, S_{is_i}$. (translation HW)

$\rightarrow F(A_1,\ldots,A_n)$ and $\rightarrow -F(A_1,\ldots,A_n)$ resp. $F(A_1,\ldots,A_n) \leftarrow$ and $-F(A_1,\ldots,A_n) \leftarrow$ are provable, if not for at least one A_j $(1 \le j \le n)$, $\rightarrow A_j$ and $\rightarrow -A_j$ resp. $A_j \leftarrow$ and $-A_j \leftarrow$ are both provable. Therefore indeed the rule schemata for introducing $-F(A_1,\ldots,A_n)$ should not be independent of the schemata (I) and (II). Now, suppose that by the schemata (I) (a), $F(A_1,\ldots,A_n)$ is provable provided that for some i, all RL-formulas in the sequence $\mathsf{X}_i = \mathsf{X}_{i1} \rightarrow (\mathsf{T}_{i1} \leftarrow \mathsf{Y}_{i1})\ldots\mathsf{X}_{is_i} \rightarrow (\mathsf{T}_{is_i} \leftarrow \mathsf{Y}_{is_i})$ are provable. Then the refutability of $F(A_1,\ldots,A_n)$ should follow from the refutability of every succession of RL-formulas X_i. If, by the schemata (I) (b), $F(A_1,\ldots,A_n)$ is provable provided that all RL-formulas $\mathsf{X}_l \rightarrow (\mathsf{T}_l \leftarrow \mathsf{Y}_l)$ $(l = 1,\ldots,j)$ are provable, then every refutation of some $\mathsf{X}_l \rightarrow (\mathsf{T}_l \leftarrow \mathsf{Y}_l)$ should be a refutation of $F(A_1,\ldots,A_n)$.

We transform these considerations into the following schemata (III) (a) and (III) (b) for introducing $-F(A_1,\ldots,A_n)$ into conclusions:

$$(III)\,(a)\quad \mathsf{W} \rightarrow -S_{11} \odot \ldots \odot -S_{1s_1} \ldots \mathsf{W} \rightarrow -S_{t1} \odot \ldots \odot -S_{ts_t} \vdash \vdash \mathsf{W} \rightarrow -F(A_1,\ldots,A_n),$$

$$-S_{11} \odot \ldots \odot -S_{1s_1} \leftarrow \mathsf{W} \ldots - S_{t1} \odot \ldots \odot -S_{ts_t} \leftarrow \mathsf{W} \vdash \vdash -F(A_1,\ldots,A_n) \leftarrow \mathsf{W};$$

$$(III)\,(b)\quad \mathsf{W} \rightarrow -S_1 \vdash \mathsf{W} \rightarrow -F(A_1,\ldots,A_n) \ldots \mathsf{W} \rightarrow -S_j \vdash \mathsf{W} \rightarrow -F(A_1,\ldots,A_n),$$
$$-S_1 \leftarrow \mathsf{W} \vdash -F(A_1,\ldots,A_n) \leftarrow \mathsf{W} \ldots - S_j \leftarrow \mathsf{W} \vdash -F(A_1,\ldots,A_n) \leftarrow \mathsf{W}.$$

Here, $\mathsf{X}_{i1} \rightarrow (T_{i1} \leftarrow \mathsf{Y}_{i1}) = S_{i1} \ldots \mathsf{X}_{is_i} \rightarrow (T_{is_i} \leftarrow \mathsf{Y}_{is_i}) = S_{is_i}$ $(1 \le i \le t)$, $\mathsf{X}_l \rightarrow (T_l \leftarrow \mathsf{Y}_l) = S_l$ $(1 \le l \le j)$, and W is an unspecified sequence of RL-formula occurrences. If F is 0-ary, we stipulate that (III) (a) and (III) (b) are not instantiated.

In the same way as the non-creativity constraint led us (in Chapter 4) from (I) (a) resp. (I) (b) to (II) (a) resp. (II) (b), it now also leads us from (III) (a) resp. (III) (b) to the following schemata (IV) (a) and (IV) (b) for introducing $-F(A_1,\ldots,A_n)$ into premises:

$$(IV)(a)\quad \mathsf{Y}_1 - S_{11} \odot \ldots \odot -S_{1s_1}\mathsf{Y}_2 \rightarrow T \vdash \mathsf{Y}_1 - F(A_1,\ldots,A_n)\mathsf{Y}_2 \rightarrow T,$$
$$\vdots$$
$$\mathsf{Y}_1 - S_{t1} \odot \ldots \odot -S_{ts_t}\mathsf{Y}_2 \rightarrow T \vdash \mathsf{Y}_1 - F(A_1,\ldots,A_n)\mathsf{Y}_2 \rightarrow T,$$
$$T \leftarrow \mathsf{Y}_1 - S_{11} \odot \ldots \odot -S_{1s_1}\mathsf{Y}_2 \vdash T \leftarrow \mathsf{Y}_1 - F(A_1,\ldots,A_n)\mathsf{Y}_2,$$
$$\vdots$$
$$T \leftarrow \mathsf{Y}_1 - S_{t1} \odot \ldots \odot -S_{ts_t}\mathsf{Y}_2 \vdash T \leftarrow \mathsf{Y}_1 - F(A_1,\ldots,A_n)\mathsf{Y}_2;$$
$$(IV)\,(b)\quad \mathsf{Y}_1 - S_1\mathsf{Y}_2 \rightarrow T \ldots \mathsf{Y}_1 - S_j\mathsf{Y}_2 \rightarrow T \vdash \mathsf{Y}_1 - F(A_1,\ldots,A_n)\mathsf{Y}_2 \rightarrow T,$$
$$T \leftarrow \mathsf{Y}_1 - S_1\mathsf{Y}_2 \ldots T \leftarrow \mathsf{Y}_1 - S_j\mathsf{Y}_2 \vdash T \leftarrow \mathsf{Y}_1 - F(A_1,\ldots,A_n);$$

where Y_1, Y_2 again are unspecified sequences of RL-formula occurrences. For 0-ary F, (IV) (a) and (IV) (b) are not instantiated.

The schemata (II) (a) resp. (II) (b), (III) (a) resp. (III) (b), and (IV) (a) resp. (IV) (b) are thus already completely determined by the schemata (I) (a) resp. (I) (b). But note that we can also assume the refutation point of view from which (I) (a) resp. (I) (b), (II) (a) resp. (II) (b), and (IV) (a) resp. (IV) (b) are already completely determined by the schemata (III) (a) resp. (III) (b). In other words, the schemata (I) and (III) mutually depend on each other.

Let C_A denote an L-formula which contains a certain occurrence of A as a subformula, and let C_B denote the result of replacing this occurrence of A in C by B. The degree of A $(d(A))$ is the number of occurrences of propositional constants and connectives in A. If $\mathsf{X} = T_1 \dots T_m$, then X_A again denotes $T_{1A} \dots T_{mA}$.

Theorem 7.6 If $\vdash A \Leftrightarrow^s B$ in **GN**$^-$ $+(I)-(IV)$, then $\vdash C_A \Leftrightarrow^s C_B$ in **GN**$^-$ $+(I)-(IV)$.

PROOF By induction on $l = d(C_A) - d(A)$. If $l = 0$, the proof is trivial. Assume that the claim holds for every $l \leq m$, and $l = m+1$. Suppose that C_A has the form $F(A_1, \dots, A_n)$, where one of the A_{kA} contains the occurrence of A in question and $d(A_{kA}) \leq l$, and $\vdash A \Leftrightarrow^s B$. By the induction hypothesis, $\vdash A_{kA} \Leftrightarrow^s A_{kB}$, and by Theorem 7.5, $\vdash$ $(\mathsf{X}_{is_i} \rightarrow (T_{is_i} \leftarrow \mathsf{Y}_{is_i}))_A \Leftrightarrow^s (\mathsf{X}_{is_i} \rightarrow (T_{is_i} \leftarrow \mathsf{Y}_{is_i}))_B$ and $\vdash (\mathsf{X}_j \rightarrow (T_j \leftarrow \mathsf{Y}_j))_A \Leftrightarrow^s$ $(\mathsf{X}_j \rightarrow (T_j \leftarrow \mathsf{Y}_j))_B$, where in each case the replacement of A by B is wrt A_k. We obtain $\vdash C_A \Leftrightarrow C_B$ as in the proof of Theorem 4.6. Assume that the rules for $-C_A$ are instantiations of (III) (a) and (IV) (a). By (ref) we have for every $i = 1, \dots, t$

$$\vdash -S_{i1_A} \odot \dots \odot -S_{is_iA} \Rightarrow -S_{i1_B} \odot \dots \odot -S_{is_iB},$$
$$\vdash -S_{i1_B} \odot \dots \odot -S_{is_iB} \Rightarrow -S_{i1_A} \odot \dots \odot -S_{is_iA}.$$

The schemata (IV) (a) give

$$\vdash -C_B \Rightarrow -S_{i1_A} \odot \dots \odot -S_{is_iA},$$
$$\vdash -C_A \Rightarrow -S_{i1_B} \odot \dots \odot -S_{is_iB}.$$

Applying (III) (a) we obtain $\vdash -C_B \Rightarrow -C_A$ and $\vdash -C_A \Rightarrow -C_B$. If the rules for $-C_A$ are instantiations of (III) (b) and (IV) (b), then by the induction hypothesis and Theorem 7.5, the schemata (III) (b) give$\vdash -(\mathsf{X}_l \rightarrow (T_l \leftarrow \mathsf{Y}_l))_A \Rightarrow -C_B$ and $\vdash -(\mathsf{X}_l \rightarrow (T_l \leftarrow \mathsf{Y}_l))_B \Rightarrow -C_A$. By (IV) (b), we obtain $\vdash -C_A \Leftrightarrow -C_B$. □

Let T_A denote an RL-formula which contains a certain occurrence of A as a subformula of a formula component of T.

Theorem 7.7 If $\vdash A \Leftrightarrow^s B$ in **GN**$^-$ $+(I)-(IV)$, then $\vdash T_A \Leftrightarrow^s T_B$ in **GN**$^-$ $+(I)-(IV)$.

PROOF By the previous two theorems. □

7.4 Functional completeness for $COSPL^-_\Delta$

In order to prove functional completeness of $COSPL^-$ resp. $COSPL^-_\Delta$ wrt the generalized Gentzen semantics resp. appropriate extensions thereof, we present sequent rules for the connectives $\sim$, /, \, $\circ$, $\wedge$, $\vee$, **t**, and $\top$. All these rules conform to the schemata (I) and (II) resp. (III) and (IV). We adopt from Chapter 4 the following rules not involving $-$:

$\underline{(\Rightarrow \mathbf{t})}$, $\underline{(\Rightarrow \mathsf{T})}$, $\underline{(\mathsf{T} \Rightarrow)}$, $\underline{(\Rightarrow /)'}$, $\underline{(/ \Rightarrow)'}$, $\underline{(\Rightarrow \backslash)'}$, $\underline{(\backslash \Rightarrow)'}$, $\underline{(\Rightarrow \circ)}$, $\underline{(\circ \Rightarrow)}$, $\underline{(\Rightarrow \wedge)}$, $\underline{(\wedge \Rightarrow)}$, and $\underline{(\rightarrow \vee)}$, $\underline{(\vee \Rightarrow)}$.

In addition to these rules we assume:

$\underline{(\Rightarrow\sim)}$ $\quad \mathsf{X} \rightarrow -T \vdash \mathsf{X} \rightarrow \sim T,$

$-T \leftarrow \mathsf{X} \vdash \sim T \leftarrow \mathsf{X};$

$\underline{(\sim\Rightarrow)}$ $\quad \mathsf{Y}_1 - T\mathsf{Y}_2 \rightarrow U \vdash \mathsf{Y}_1 \sim T\mathsf{Y}_2 \rightarrow U,$

$U \leftarrow \mathsf{Y}_1 - T\mathsf{Y}_2 \vdash U \leftarrow \mathsf{Y}_1 \sim T\mathsf{Y}_2;$

$\underline{(\Rightarrow -/)'}$ $\quad \mathsf{X} \rightarrow -(U \leftarrow T) \vdash \mathsf{X} \rightarrow -(U/T),$

$-(U \leftarrow T) \leftarrow \mathsf{X} \vdash -(U/T) \leftarrow \mathsf{X};$

$\underline{(\Rightarrow -\backslash)'}$ $\quad \mathsf{X} \rightarrow -(T \rightarrow U) \vdash \mathsf{X} \rightarrow -(T \backslash U),$

$-(T \rightarrow U) \leftarrow \mathsf{X} \vdash -(T \backslash U) \leftarrow \mathsf{X};$

$\underline{(-/ \Rightarrow)'}$ $\quad \mathsf{Y}_1 - (U \leftarrow T)\mathsf{Y}_2 \rightarrow T_1 \vdash \mathsf{Y}_1 - (U/T)\mathsf{Y}_2 \rightarrow T_1,$

$T_1 \leftarrow \mathsf{Y}_1 - (U \leftarrow T)\mathsf{Y}_2 \vdash T_1 \leftarrow \mathsf{Y}_1 - (U/T)\mathsf{Y}_2;$

$\underline{(-\backslash \Rightarrow)}$ $\quad \mathsf{Y}_1 - (T \rightarrow U)\mathsf{Y}_2 \rightarrow T_1 \vdash \mathsf{Y}_1 - (T \backslash U)\mathsf{Y}_2 \rightarrow T_1,$

$T_1 \leftarrow \mathsf{Y}_1 - (T \rightarrow U)\mathsf{Y}_2 \vdash T_1 \leftarrow \mathsf{Y}_1 - (T \backslash U)\mathsf{Y}_2;$

$\underline{(\Rightarrow -\wedge)}$ $\quad \mathsf{X} \rightarrow -T \vdash \mathsf{X} \rightarrow -(T \wedge U),$

$\mathsf{X} \rightarrow -U \vdash \mathsf{X} \rightarrow -(T \wedge U),$

$-T \leftarrow \mathsf{X} \vdash -(T \wedge U) \leftarrow \mathsf{X},$

$-U \leftarrow \mathsf{X} \vdash -(T \wedge U) \leftarrow \mathsf{X};$

$\underline{(-\wedge \Rightarrow)}$ $\quad (\mathsf{Y}_1 - T\mathsf{Y}_2 \rightarrow T_1)(\mathsf{Y}_1 - U\mathsf{Y}_2 \rightarrow T_1) \vdash \mathsf{Y}_1 - (T \wedge U)\mathsf{Y}_2 \rightarrow T_1,$

$(T_1 \leftarrow \mathsf{Y}_1 - T\mathsf{Y}_2)(T_1 \leftarrow \mathsf{Y}_1 - U\mathsf{Y}_2) \vdash T_1 \leftarrow \mathsf{Y}_1 - (T \wedge U)\mathsf{Y}_2;$

$\underline{(\Rightarrow -\circ)'}$ $\quad \mathsf{X} \rightarrow (-T \odot -U) \vdash \mathsf{X} \rightarrow -(T \circ U),$

$(-T \odot -U) \leftarrow \mathsf{X} \vdash -(T \circ U) \leftarrow \mathsf{X};$

$\underline{(-\circ \Rightarrow)'}$ $\quad \mathsf{Y}_1(-T \odot -U)\mathsf{Y}_2 \rightarrow T_1 \vdash \mathsf{Y}_1 - (T \circ U)\mathsf{Y}_2 \rightarrow T_1,$

$T_1 \leftarrow \mathsf{Y}_1(-T \odot -U)\mathsf{Y}_2 \vdash T_1 \leftarrow \mathsf{Y}_1 - (T \circ U)\mathsf{Y}_2;$

$\underline{(\Rightarrow -\vee)}$ $\quad (\mathsf{X} \rightarrow -T)(\mathsf{X} \rightarrow -U) \vdash \mathsf{X} \rightarrow -(T \vee U),$

$(-T \leftarrow \mathsf{X})(-U \leftarrow \mathsf{X}) \vdash -(T \vee U) \leftarrow \mathsf{X};$

$\underline{(-\vee \Rightarrow)}$ $\quad \mathsf{Y}_1 - T\mathsf{Y}_2 \rightarrow T_1 \vdash \mathsf{Y}_1 - (T \vee U)\mathsf{Y}_2 \rightarrow T_1,$

$\mathsf{Y}_1 - U\mathsf{Y}_2 \rightarrow T_1 \vdash \mathsf{Y}_1 - (T \vee U)\mathsf{Y}_2 \rightarrow T_1,$

$T_1 \leftarrow \mathsf{Y}_1 - T\mathsf{Y}_2 \vdash T_1 \leftarrow \mathsf{Y}_1 - (T \vee U)\mathsf{Y}_2,$

$T_1 \leftarrow \mathsf{Y}_1 - U\mathsf{Y}_2 \vdash T_1 \leftarrow \mathsf{Y}_1 - (T \vee U)\mathsf{Y}_2;$

$\underline{(\Rightarrow - \sim)}$ $\quad \mathsf{X} \rightarrow - - T \vdash \mathsf{X} \rightarrow - \sim T,$

$- - T \leftarrow \mathsf{X} \vdash - \sim T \leftarrow \mathsf{X};$

$$\begin{array}{ll}\underline{(-\sim\Rightarrow)} & \mathsf{Y}_1 - -T\mathsf{Y}_2 \to U \vdash \mathsf{Y}_1 - \sim T\mathsf{Y}_2 \to U, \\ & U \leftarrow \mathsf{Y}_1 - -T\mathsf{Y}_2 \vdash U \leftarrow \mathsf{Y}_1 - \sim T\mathsf{Y}_2.\end{array}$$

From these rules it is clear that $\vdash (T \odot U) \Leftrightarrow^s (T \circ U)$, $\vdash -T \Leftrightarrow^s \sim T$, $\vdash (U \leftarrow T) \Leftrightarrow^s (U/T)$, and $\vdash (T \to U) \Leftrightarrow^s (T \setminus U)$.

In Chapter 4 we have remarked that $\underline{(\Rightarrow /)'}$ resp. $\underline{(\Rightarrow \setminus)'}$ is equivalent to

$$\begin{array}{ll}\underline{(\Rightarrow /)} & \mathsf{X}T \to U \vdash \mathsf{X} \to (U/T), \\ & U \leftarrow \mathsf{X}T \vdash (U/T) \leftarrow \mathsf{X}; \text{ resp.} \\ \underline{(\Rightarrow \setminus)} & T\mathsf{X} \to U \vdash \mathsf{X} \to (T \setminus U), \\ & U \leftarrow T\mathsf{X} \vdash (T \setminus U) \leftarrow \mathsf{X};\end{array}$$

and that $\underline{(/ \Rightarrow)'}$ resp. $\underline{(\setminus \Rightarrow)'}$ is equivalent to

$$\begin{array}{ll}\underline{(/ \Rightarrow)} & \mathsf{X} \to (U/T) \vdash \mathsf{X}T \to U, \\ & (U/T) \leftarrow \mathsf{X} \vdash U \leftarrow \mathsf{X}T; \text{ resp.} \\ \underline{(\setminus \Rightarrow)} & \mathsf{X} \to (T \setminus U) \vdash T\mathsf{X} \to U, \\ & (T \setminus U) \leftarrow \mathsf{X} \vdash U \leftarrow T\mathsf{X}.\end{array}$$

It can also easily be seen that $\underline{(\Rightarrow -/)'}$ resp. $\underline{(\Rightarrow -\setminus)'}$ is equivalent to

$$\begin{array}{ll}\underline{(\Rightarrow -/)} & (\mathsf{X} \to -U)(\mathsf{Y} \to T) \vdash \mathsf{XY} \to -(U/T), \\ & (-U \leftarrow \mathsf{X})(T \leftarrow \mathsf{Y}) \vdash -(U/T) \leftarrow \mathsf{XY}; \quad \text{resp.} \\ \underline{(\Rightarrow -\setminus)} & (\mathsf{X} \to T)(\mathsf{Y} \to -U) \vdash \mathsf{XY} \to -(T \setminus U), \\ & (T \leftarrow \mathsf{X})(-U \leftarrow \mathsf{Y}) \vdash -(T \setminus U) \leftarrow \mathsf{XY}.\end{array}$$

See e.g. the following derivation:

$$\dfrac{\dfrac{\mathsf{X} \to -(U \leftarrow T) \quad \dfrac{\dfrac{-U \to -U \quad T \to T}{-UT \to (-U \circ T)}}{-(U \leftarrow T) \to (-U \circ T)}}{\mathsf{X} \to (-U \circ T)} \qquad \dfrac{\dfrac{-U \to -U \quad T \to T}{(-UT) \to -(U/T)}}{(-U \circ T) \to -(U/T)}}{\mathsf{X} \to -(U/T).}$$

Moreover, $\underline{(-/ \Rightarrow)'}$ resp. $\underline{(-\setminus \Rightarrow)'}$ is equivalent to

$$\begin{array}{ll}\underline{(-/ \Rightarrow)} & \mathsf{Y}_1 - UT\mathsf{Y}_2 \to T_1 \vdash \mathsf{Y}_1 - (U/T)\mathsf{Y}_2 \to T_1, \\ & T_1 \leftarrow \mathsf{Y}_1 - UT\mathsf{Y}_2 \vdash T_1 \leftarrow \mathsf{Y}_1 - (U/T)\mathsf{Y}_2; \quad \text{resp.} \\ \underline{(-\setminus \Rightarrow)} & \mathsf{Y}_1 T - U\mathsf{Y}_2 \to T_1 \vdash \mathsf{Y}_1 - (T \setminus U)\mathsf{Y}_2 \to T_1, \\ & T_1 \leftarrow \mathsf{Y}_1 T - U\mathsf{Y}_2 \vdash T_1 \leftarrow \mathsf{Y}_1 - (T \setminus U)\mathsf{Y}_2;\end{array}$$

and $\underline{(\Rightarrow -\circ)'}$ resp. $\underline{(-\circ \Rightarrow)'}$ is equivalent to

$$\begin{array}{ll}\underline{(\Rightarrow -\circ)} & (\mathsf{X} \to -T_1)(\mathsf{Y} \to -T_2) \vdash \mathsf{XY} \to -(T_1 \circ T_2); \\ & (-T_1 \leftarrow \mathsf{X})(-T_2 \leftarrow \mathsf{Y}) \vdash -(T_1 \circ T_2) \leftarrow \mathsf{XY}; \\ \underline{(-\circ \Rightarrow)} & \mathsf{X} - T - U\mathsf{Y} \to T_1 \vdash \mathsf{X} - (T \circ U)\mathsf{Y} \to T_1; \\ & T_1 \leftarrow \mathsf{X} - T - U\mathsf{Y} \vdash T_1 \leftarrow \mathsf{X} - (T \circ U)\mathsf{Y}.\end{array}$$

Next, we assign to each $R\mathsf{L}$-formula T one formula $\overline{T}$ with $Rd(\overline{T}) = 0$ by stipulating:

if T is an L-formula, then $\overline{T} = T$;

if $T = -U$, then $\overline{T} = \sim \overline{T}$;

if $T = T_1 \odot T_2$, then $\overline{T} = \overline{T_1} \circ \overline{T_2}$;

if $\mathsf{X} = T_1 \ldots T_n$, then $\overline{\mathsf{X}} = \overline{T_1} \circ \ldots \circ \overline{T_n}$;

if $T = \mathsf{X} \rightarrow U$, then $\overline{T} = \overline{\mathsf{X}} \setminus \overline{U}$;

if $T = U \leftarrow \mathsf{X}$, then $\overline{T} = \overline{U} / \overline{\mathsf{X}}$;

if $T = \; \rightarrow U$, then $\overline{T} = \top \setminus \overline{U}$;

if $T = U \leftarrow \;$, then $\overline{T} = \overline{U} / \top$.

Theorem 7.8 $\vdash T \Leftrightarrow^s \overline{T}$ in $\mathbf{GN}^-$ $+(I) - (IV)$.

PROOF By induction on $Rd(T)$. If $Rd(T) = 0$, the claim is trivial. Suppose that the claim holds for every $l \leq m$, and $l = m+1$. By the induction hypothesis and the previous theorem we have $\vdash (T_1 \ldots T_n \rightarrow U) \Leftrightarrow^s (\overline{T_1} \ldots \overline{T_n} \rightarrow \overline{U})$ resp. $\vdash (U \leftarrow T_1 \ldots T_n) \Leftrightarrow^s (\overline{U} \leftarrow \overline{T_1} \ldots \overline{T_n})$ resp. $\vdash (\rightarrow T) \Leftrightarrow^s (\rightarrow \overline{T})$ resp. $\vdash (T \leftarrow) \Leftrightarrow^s (\overline{T} \leftarrow)$. Now, by $(\rightarrow \circ)$, $(\circ \rightarrow)$, (tra), and the rules of $\mathbf{GN}^-$ referring to $-$, we obtain $\vdash (\overline{T_1} \ldots \overline{T_n} \rightarrow \overline{U}) \Leftrightarrow^s (\overline{T_1} \circ \ldots \circ \overline{T_n} \rightarrow \overline{U})$ and $\vdash (\overline{U} \leftarrow \overline{T_1} \ldots \overline{T_n}) \Leftrightarrow^s (\overline{U} \leftarrow \overline{T_1} \circ \ldots \circ \overline{T_n})$. Since $\vdash (U \leftarrow T) \Leftrightarrow^s (U/T)$ and $\vdash (T \rightarrow U) \Leftrightarrow^s (T \setminus U)$, $\vdash (\overline{T_1} \ldots \overline{T_n} \rightarrow \overline{U}) \Leftrightarrow^s (\overline{T_1} \circ \ldots \circ \overline{T_n} \setminus \overline{U})$ and $\vdash (\overline{U} \leftarrow \overline{T_1} \ldots \overline{T_n}) \Leftrightarrow^s (\overline{U} \,/\, \overline{T_1} \circ \ldots \circ \overline{T_n})$. Thus, $\vdash (T_1 \ldots T_n \rightarrow U) \Leftrightarrow^s \overline{(T_1 \ldots T_n \rightarrow U)}$ and $\vdash (U \leftarrow T_1 \ldots T_n) \Leftrightarrow^s \overline{(U \leftarrow T_1 \ldots T_n)}$. Since $\vdash -T \Leftrightarrow^s \sim T$, we also have $\vdash -(T_1 \ldots T_n \rightarrow U) \Leftrightarrow^s \overline{-(T_1 \ldots T_n \rightarrow U)}$ and $\vdash -(U \leftarrow T_1 \ldots T_n) \Leftrightarrow^s \overline{-(U \leftarrow T_1 \ldots T_n)}$. Moreover, by the induction hypothesis and the previous theorem $\vdash (T \odot U) \Leftrightarrow^s (\overline{T} \odot \overline{U})$. Since $(T \odot U) \Leftrightarrow^s (T \circ U)$, we have $(T \odot U) \Leftrightarrow^s \overline{(T \odot U)}$. Finally, since $\vdash -T \Leftrightarrow^s \sim T$, we also have $-(T \odot U) \Leftrightarrow^s \overline{-(T \odot U)}$. □

Theorem 7.9 $\{\sim, /, \setminus, \wedge, \circ, \vee, \mathbf{t}, \top\}$ is functionally complete wrt the generalized Gentzen semantics.

PROOF As in the proof of Theorem 4.9, we can show that, if $n > 0$, $\vdash F(A_1, \ldots, A_n) \Leftrightarrow B$, where $B = \overline{\mathsf{X}_1} \vee \ldots \vee \overline{\mathsf{X}_t}$, if the rules for $F(A_1, \ldots, A_n)$ are instantiations of (I) (a) and (II) (a) resp. $B = \overline{S_1} \wedge \ldots \wedge \overline{S_j}$, if the rules for $F(A_1, \ldots, A_n)$ are instantiations of (I) (b) and (II) (b). If $n = 0$, then $F = \top$ or $F = \mathbf{t}$. We are done, if we can show that $\vdash -B \Leftrightarrow -F(A_1, \ldots, A_n)$, for $n > 0$. Suppose that $-F(A_1, \ldots, A_n)$ is defined by instantiations of the schemata (III) (a) and (IV) (a). By (ref), $(\odot \Uparrow)$, $(-\odot \rightarrow)$, $(\leftarrow -\odot)$, the previous theorem, and the fact that $\sim T \Leftrightarrow^s -T$, we have $\vdash -\overline{\mathsf{X}_i} \Rightarrow -S_{i1} \odot \ldots \odot -S_{is_i}$. Applying $(\wedge \Rightarrow)$, we obtain

$$\vdash -\overline{\mathsf{X}_1} \wedge \ldots \wedge -\overline{\mathsf{X}_t} \Rightarrow -S_{i1} \odot \ldots \odot -S_{is_i}.$$

The schemata (III) (a) give

$$\vdash -\overline{\mathsf{X}_1} \wedge \ldots \wedge -\overline{\mathsf{X}_t} \Rightarrow -F(A_1, \ldots, A_n).$$

and from this we easily obtain

$$\vdash -(\overline{X_1} \vee \ldots \vee \overline{X_t}) \Rightarrow -F(A_1, \ldots, A_n).$$

By (ref), $(\odot \rightarrow)$, $(\leftarrow \odot)$, $(\rightarrow -\odot)$, $(-\odot \leftarrow)$, (tra), the previous theorem, and the fact that $\sim T \Leftrightarrow^s -T$, we have $\vdash -S_{i1} \odot \ldots \odot -S_{is_i} \Rightarrow -\overline{X_i}$. The schemata (IV) (a) give $\vdash -F(A_1, \ldots, A_n) \Rightarrow -\overline{X_i}$. Applying $\underline{(\Rightarrow \wedge)}$, we obtain

$$\vdash -F(A_1, \ldots, A_n) \Rightarrow -\overline{X_1} \wedge \ldots \wedge -\overline{X_t}.$$

From the latter we readily obtain

$$\vdash -F(A_1, \ldots, A_n) \Rightarrow -(\overline{X_1} \vee \ldots \vee \overline{X_t}).$$

Assume now that the rules for $-F(A_1, \ldots, A_n)$ are instantiations of the schemata (III) (b) and (IV) (b). By (ref), the previous theorem, the fact that $\sim T \Leftrightarrow^s -T$, and (tra), $\vdash -\overline{S_l} \Rightarrow -S_l$. The schemata (III) (b) give $\vdash -\overline{S_l} \Rightarrow -F(A_1, \ldots, A_n)$. Applying $\underline{(\vee \Rightarrow)}$, we may conclude that

$$\vdash -\overline{S_1} \vee \ldots \vee -\overline{S_t} \Rightarrow -F(A_1, \ldots, A_n);$$

and from this it is an easy matter to deduce

$$\vdash -(\overline{S_1} \wedge \ldots \wedge \overline{S_t}) \Rightarrow -F(A_1, \ldots, A_n).$$

Moreover, clearly $\vdash -S_l \Rightarrow -\overline{S_l}$. Hence, by $\underline{(\Rightarrow \vee)}$, $\vdash -S_l \Rightarrow -\overline{S_1} \vee \ldots \vee -\overline{S_j}$. By the schemata (IV) (b), we can prove $-F(A_1, \ldots, A_n) \Rightarrow -\overline{S_1} \vee \ldots \vee -\overline{S_j}$, from which $-F(A_1, \ldots, A_n) \Rightarrow -(\overline{S_1} \wedge \ldots \wedge \overline{S_j})$ is derivable. □

It is not difficult to show that the generalized Gentzen semantics characterizes $COSPL^-$.

Theorem 7.10 $\{/, \backslash, \wedge, \circ, \vee, \top, \mathbf{t}, \sim\}$ is functionally complete for $COSPL^-$.

PROOF If (i) RL-formulas and the premises and conclusions of $\underline{(\Rightarrow \mathbf{t})}$, $\underline{(\Rightarrow \top)}$, $\underline{(\top \Rightarrow)}$, $\underline{(\Rightarrow F)}$, $\underline{(F \Rightarrow)}$, $\underline{(\Rightarrow -F)}$, and $\underline{(-F \Rightarrow)}$ $(F \in \{/, \backslash, \circ, \wedge, \vee, \sim\})$ are restricted to sequents only, (ii) $T \leftarrow T_n \ldots T_1$ is read as its mirror-image $T_1 \ldots T_n \rightarrow T$, and $-T$ is read as $\sim T$, then the resulting calculus is equivalent to $COSPL^-$ in the sense that both systems have the same set of provable sequents, as a comparison of the systems immediately reveals. □

This functional completeness result can straightforwardly be extended from $COSPL^-$ to $COSPL^-_\Delta$ in the same way as the functional completeness result for $ISPL$ in Chapter 4 has been extended to $ISPL_\Delta$, viz. by enriching $\mathbf{GN}^-$ by higher-level versions of the structural inference rules in Δ, using both sequent arrows $\rightarrow$ and $\leftarrow$.

Corollary 7.11 $\{/, \backslash, \wedge, \circ, \vee, \top, \mathbf{t}, \sim\}$ is functionally complete for $COSPL^-_\Delta$.[5]

[5] We do not consider the intuitionistic minimal negations $\neg^r$, $\neg^l$, which are definable in every system $COSPL^-_\Delta$, as essential to these logics, since, in contrast to $MSPL_\Delta$, without $\neg^r$ and $\neg^l$, $COSPL^-_\Delta$ is not void of any 'official' negation. If nevertheless $\neg^r$ and $\neg^l$ are regarded as integral to $COSPL^-_\Delta$, $\bot$ should be assumed as a designated propositional variable of the underlying higher-level Gentzen calculus and should be added to the set $\{/, \backslash, \wedge, \circ, \vee, \top, \mathbf{t}, \sim\}$.

7.5 Functional completeness for $COSPL_\Delta$

Let **GN** denote the result of adding the rules $(\bot \Rightarrow)$ (i.e. $\mathsf{X}\bot\mathsf{Y} \Rightarrow T$), and

$$(\Rightarrow -\bot) \;\vdash \mathsf{X} \Rightarrow -\bot$$

to $\mathbf{GN}^-$. Moreover, let for $n = 0$ the rule schemata (III) (a) and (III) (b) both be instantiated by $\vdash \mathsf{X} - F\mathsf{Y} \Rightarrow T$.

In complete analogy to the above argument for $COSPL_\Delta^-$ we obtain a functional completeness result for $COSPL_\Delta$ wrt to this extended generalized Gentzen semantics. Remember that in $COSPL_\Delta$ one can prove (i) $\sim \bot \rightleftharpoons^+ \mathbf{t}$ and (ii) $\mathbf{t} \rightleftharpoons^+ (\bot \setminus \bot)$, $\sim \mathbf{t} \rightleftharpoons^+ \bot$. Therefore we obtain the following

Theorem 7.12 $\{/, \setminus, \wedge, \circ, \vee, \top, \mathbf{t}, \sim\}$, and $\{/, \setminus, \wedge, \circ, \vee, \top, \bot, \sim\}$ are functionally complete for $COSPL_\Delta$.

Since in $COSPL_\Delta^-$ and $COSPL_\Delta$ we have $\vdash (A \wedge B) \rightleftharpoons\sim (\sim A \vee \sim B)$ as well as $\vdash (A \vee B) \rightleftharpoons\sim (\sim A \wedge \sim B)$, we may drop either $\vee$ or $\wedge$ (but not both) from the sets of connectives wrt which $COSPL_\Delta^-$ and $COSPL_\Delta$ have been shown to be functionally complete and again obtain functionally complete sets of connectives.

7.6 Digression: Negation in Categorial Grammar

An issue which we have completely neglected in our discussion in Chapter 4 of additional connectives for the Lambek Calculus is the problem of negative linguistic information. Our functional completeness result for $PSPL$ has been obtained from the corresponding result for $ISPL$, i.e. from $PSPL$ augmented by an intuitionistic falsum $\bot$, intuitionistic negations $\neg^r$ and $\neg^l$ being defined by $\neg^r A \stackrel{def}{=} (\bot/A)$, $\neg^l A \stackrel{def}{=} (A \setminus \bot)$. These intuitionistic negations, however, do not seem to admit of a linguisitc interpretation. One might e.g. think of interpreting $\bot$ as "ungrammatical", but unfortunately it is simply not the case that a linguistic item is, say, a non-sentence, just in case its combination with a sentence results in an ungrammatical string.

On the other hand, the usual treatment of negative linguistic information in Categorial Grammar is not quite satisfactory. What one can find in the literature is the idea of a negation as failure, which is tacitly based on a certain kind of *closed world assumption* and which is not reflected by a negation sign in the object language. Lambek [1958] e.g. describes his aim as "to obtain an effective rule (or algorithm) for distinguishing sentences from nonsentences, which works not only for the formal languages of interest to the mathematical logician, but also for natural languages such as English, or at least for fragments of such languages" [p. 154]. Thus, a given string of linguistic items does not belong to type s, if for every type assignment to the given items the effective algorithm for distinguishing sentences from nonsentences does not terminate with an s, i.e., if all attempts to derive sentencehood fail. But clearly failure to derive sentencehood differs from deriving nonsentencehood. Even if one is working with a decidable syntactic calculus, negation as failure to derive makes sense in the categorial analysis of language only against the background of the following, usually tacit closed world assumption:

CWA Every linguistic item under consideration has already been completely categorized in the sense that every syntactic type in which the item may occur can be derived from an assigned type by means of the underlying syntactic calculus.

If incomplete, partial knowledge of a language's categorial structure (at the level of assignment functions) is not excluded, one has to reckon with the possibility that there are sentences among those strings which are not recognized as sentences by the terminating algorithm. Thus, the negation as failure principle no longer seems to be justified. Like intuitionistic negation also neither classical nor strong, constructive negation $\sim$ allow for a linguisitc interpretation. In the classical resp. constructive case we would e.g. have that $\neg(s/s)$ resp. $\sim (s/s)$ is interderivable with $(\neg s \circ s)$ resp. $(\sim s \circ s)$. However, if a linguistic item is not an adverb, this doesn't mean that one is dealing with a string consisting of a non-sentence followed by a sentence. Thus, what is an appropriate negation operation for Categorial Grammar still has to be investigated.

Chapter 8

The constructive typed λ-calculus λ^c and formulas-as-types for $\mathbf{N}^-$

In Chapter 5 we have studied the formulas-as-types notion of construction for substructural subsystems of intuitionistic propositional logic IPL. There we have restricted our attention almost entirely to syntactic aspects of the encoding of terms by proofs, and vice versa. We may consider this detailed investigation as exemplary for what happens to the idea of formulas-as-types in the absence of structural inference rules and therefore in the present chapter focus our interest on the *semantics* of the typed λ-calculus which we want to associate with $COSPL^-_{\{\mathbf{P},\mathbf{C},\mathbf{M}\}}$ alias $\mathbf{N}^-$. We shall introduce the constructive typed λ-caluclus λ^c and prove completeness of λ^c wrt what will be called ***full constructive type structures over infinite sets***. This result extends H. Friedman's [1975] completeness proof for $\lambda_\supset$ wrt to so-called full type structures over infinite sets. As a kind of justification for calling λ^c *constructive*, we shall then encode proofs in $\mathbf{N}^-$ by terms from λ^c's set of typed terms, and conversely.

8.1 The typed λ-calculus λ^c

8.1.1 The syntax of λ^c

The set of type symbols (or just types) is the propositional language in the connectives $\sim$, $\wedge$, $\vee$, and $\supset$ based on $PROP \cup \{\bot\}$. The set VAR of term variables is defined as $\{v_i^A \mid 0 < i \in \omega,\ A \text{ is a type}\}$.

Definition 8.1 The set $TERM$ of terms is the smallest set Γ such that

$VAR \subseteq \Gamma$;

if M^A, $N^B \in \Gamma$, then $<M, N>^{(A\wedge B)} \in \Gamma$;

if $M^A \in \Gamma$, then $K^0_{A,B}(M)^{(A\vee B)} \in \Gamma$;

if $M^B \in \Gamma$, then $K^1_{A,B}(M)^{(A\vee B)} \in \Gamma$;

if $M^B \in \Gamma$, $x^A \in VAR$, then $(\lambda x M)^{(A\supset B)} \in \Gamma$;

if $M^{(A \wedge B)} \in \Gamma$, then $(M)_0^A$, $(M)_1^B \in \Gamma$;[1]

if $M^{(A \supset B)}$, $N^{(B \supset C)} \in \Gamma$, then $[M, N]^{((A \vee B) \supset C)} \in \Gamma$;

if $M^{(A \supset B)}$, $N^A \in \Gamma$, then $(M, N)^B \in \Gamma$;

$M^{\sim\sim A} \in \Gamma$ iff $M^A \in \Gamma$;

$M^{\sim(A \wedge B)} \in \Gamma$ iff $M^{(\sim A \vee \sim B)} \in \Gamma$;

$M^{\sim(A \vee B)} \in \Gamma$ iff $M^{(\sim A \wedge \sim B)} \in \Gamma$;

$M^{\sim(A \supset B)} \in \Gamma$ iff $M^{(A \wedge \sim B)} \in \Gamma$.

We shall denote term variables by x, y, z, x_1, x_2, etc. and terms by M, N, G, M_1, M_2, etcetera. We say that M^A is a term of type A.

Definition 8.1 has as a non-standard consequence the *non-unique-typedness* of terms. If M is a term of more than one type, then these types are related by the provable acceptance equivalences

$$\sim\sim A \rightleftharpoons^+ A, \qquad \sim(A \supset B) \rightleftharpoons^+ (A \wedge \sim B),$$
$$\sim(A \wedge B) \rightleftharpoons^+ (\sim A \vee \sim B), \quad \sim(A \vee B) \rightleftharpoons^+ (\sim A \wedge \sim B),$$

where $A \rightleftharpoons^+ B$ here abbreviates $(A \supset B) \wedge (B \supset A)$. Allowing for terms of multiple type has to do with atomicity of negation as a consequence of regarding the negative as a counterpart of the positve. Although in λ^c we have in addition to lambda-abstraction and application also term forming operations in order to deal with $\wedge$ and $\vee$, there are no operations for $\sim$. We deal with constructive negation not by introducing further operations but instead by looking differently at the λ-terms; they may occur in more than one type.

Definition 8.2 $FV(M)$, the set of free variables of M, is inductively defined as follows:

$FV(x) = \{x\}$;

$FV(< M, N >) = FV(M) \cup FV(N)$;

$FV((M)_i) = FV(M)$, $i = 0, 1$;

$FV(K^i(M)) = FV(M)$, $i = 0, 1$;

$FV(\lambda x M) = FV(M) - \{x\}$;

$FV([G_1, G_2]) = FV(G_1) \cup FV(G_2)$;

$FV((M, N)) = FV(M) \cup FV(N)$.

The variable x is bound in M ($x \in BV(M)$) iff $x \notin FV(M)$.

[1] As in Chapter 5, we allow projections of terms which are not pairs. We need such projections for the encoding of proofs with $\wedge$-introductions on the lhs of $\rightarrow$; see Section 8.2.

Definition 8.3 $M[x^B := N^B]$, the result of substituting N for each free occurrence of x in M, is inductively defined by:

$x^B[x^B := N^B] = N$;

$x^A[x^B := N^B] = x^A$;

$< M, G > [x := N] =< M[x := N], G[x := N] >$;

$K^i(M)[x := N] = K^i(M[x := N]),\ i = 0, 1$;

$(\lambda y M)[x := N] = (\lambda y M[x := N])$, if $y \notin FV(N)$;

$(M)_i[x := N] = (M[x := N])_i,\ i = 0, 1$;

$[G_1, G_2][x := N] = [G_1[x := N], G_2[x := N]]$;

$(M, G)[x := N] = (M[x := N], G[x := N])$.

Definition 8.4 The axiom-schemata and rules of λ^c are:

(α)	$(\lambda x^A M) = (\lambda y^A M[x := y])$, if $y \notin FV(M) \cup BV(M)$;
(β)	$((\lambda x M)N) = M[x := N]$, if $BV(M) \cap FV(N) = \emptyset$;
(η)	$(\lambda x(Mx)) = M$, if $x \notin FV(M)$;
$< 0 >$	$(< M, N >)_0 = M$;
$< 1 >$	$(< M, N >)_1 = N$;
$< sur >$	$< (M)_0, (M)_1 >= M$;
(0)	$([M^{(A \supset C)}, N^{(B \supset C)}], K^0_{A,B}(G)) = (M, G)$;
(1)	$([M^{(A \supset C)}, N^{(B \supset C)}], K^1_{A,B}(G)) = (N, G)$;
$(0, 1)$	$K^0_{A,B}(M) = K^1_{A,B}(M) = K^0_{B,A}(M) = K^1_{B,A}(M)$, if $M^A = M^B$;
(ϑ_1)	$([(\lambda x^A(M^{(A \vee B) \supset C} K^0_{A,B}(x))), (\lambda y^B(M^{(A \vee B) \supset C} K^1_{A,B}(y)))], G) = (M, G)$, if $x, y \notin FV(M)$;
(ϑ_2)	$([(\lambda x^A K^0_{A,B}(x)), (\lambda y^B K^1_{A,B}(y))], G) = G$;
$(\sim)$	$M^{\sim\sim A} = M^A$; $M^{\sim(A \wedge B)} = M^{(\sim A \vee \sim B)}$; $M^{\sim(A \vee B)} = M^{(\sim A \wedge \sim B)}$; $M^{\sim(A \supset B)} = M^{(A \wedge \sim B)}$;
1	$M^A = M^A$, $M = N \vdash N = M$; $M = N\ N = G \vdash M = G$;
2	$M = N \vdash < G, M >=< G, N >$; $M = N \vdash < M, G >=< N, G >$; $M = N \vdash (G, M) = (G, N)$; $M = N \vdash (M, G) = (N, G)$; $M = N \vdash (\lambda x M) = (\lambda x N)$; $M = N \vdash (M)_i = (N)_i,\ i = 0, 1$; $M = N \vdash K^i(M) = K^i(N),\ i = 0, 1$; $M = N \vdash [G, M] = [G, N]$; $M = N \vdash [M, G] = [N, G]$.

Let $\gg$ denote the reduction relation on $TERM$ induced by (the direction from left to right of) (α), (β), (η), $<0>$, $<1>$, $<sur>$, (0), (1), (ϑ_1), (ϑ_2), $(\sim)$, and (both directions of) (0, 1).

QUESTION (i) Is every M strongly normalizable wrt $\gg$?
(ii) Is $\gg$ Church-Rosser?

8.1.2 Models for λ^c

Definition 8.5 $\mathcal{F} =< \{\mathsf{D}^A\}, \{\mathsf{AP}_{A,B}\}, \{\mathsf{PRO}^0_{A,B}\}, \{\mathsf{PRO}^1_{A,B}\}, \{\mathsf{PAIR}_{A,B}\}, \{\mathsf{DIS}_{A,B,C}\}, \{\mathsf{K}^0_{A,B}\}, \{\mathsf{K}^1_{A,B}\} >$ is called a type structure frame (or just a frame) iff

- D^A is a non-empty set, for each type A;
- $\mathsf{K}^0_{A,B} : \mathsf{D}^A \longrightarrow \mathsf{D}^{(A\vee B)}$,
  $\mathsf{K}^1_{A,B} : \mathsf{D}^B \longrightarrow \mathsf{D}^{(A\vee B)}$,
  $\mathsf{AP}_{A,B} : \mathsf{D}^{(A\supset B)} \times \mathsf{D}^A \longrightarrow \mathsf{D}^B$,
  $\mathsf{PRO}^0_{A,B} : \mathsf{D}^{(A\wedge B)} \longrightarrow \mathsf{D}^A$,
  $\mathsf{PRO}^1_{A,B} : \mathsf{D}^{(A\wedge B)} \longrightarrow \mathsf{D}^B$,
  $\mathsf{PAIR}_{A,B} : \mathsf{D}^A \times \mathsf{D}^B \longrightarrow \mathsf{D}^{(A\wedge B)}$,
  $\mathsf{DIS}_{A,B,C} : \mathsf{D}^{(A\supset C)} \times \mathsf{D}^{(B\supset C)} \longrightarrow \mathsf{D}^{((A\vee B)\supset C)}$, for all types A, B, C;
- (extensionality) if $a, b \in \mathsf{D}^{(A\supset B)}$ and $(\forall c \in \mathsf{D}^A)(\mathsf{AP}_{A,B}(a,c) = \mathsf{AP}_{A,B}(b,c))$, then $a = b$;
- (red $\wedge$) for all $a \in \mathsf{D}^A$, $b \in \mathsf{D}^B$:
  (i) $\mathsf{PRO}^0_{A,B}(\mathsf{PAIR}_{A,B}(a,b)) = a$,
  (ii) $\mathsf{PRO}^1_{A,B}(\mathsf{PAIR}_{A,B}(a,b)) = b$;
- (red $\vee$) for all $a \in \mathsf{D}^{(A\supset)C}$, $b \in \mathsf{D}^{(B\supset C)}$, $a_1 \in \mathsf{D}^A$, $b_1 \in \mathsf{D}^B$:
  (i) $\mathsf{AP}(\mathsf{DIS}(a,b), \mathsf{K}^0_{A,B}(a_1)) = \mathsf{AP}(a, a_1)$;
  (ii) $\mathsf{AP}(\mathsf{DIS}(a,b), \mathsf{K}^1_{A,B}(b_1)) = \mathsf{AP}(b, b_1)$;
- $\mathsf{K}^0_{A,B}(a) = \mathsf{K}^1_{A,B}(a) = \mathsf{K}^0_{B,A}(a) = \mathsf{K}^1_{B,A}(a), \forall a \in \mathsf{D}^A$, if $\mathsf{D}^A = \mathsf{D}^B$.

Definition 8.6 Let $\mathcal{F} = < \{\mathsf{D}^A\}, \{\mathsf{AP}_{A,B}\}, \{\mathsf{PRO}^0_{A,B}\}, \{\mathsf{PRO}^0_{A,B}\}, \{\mathsf{PAIR}_{A,B}\}, \{\mathsf{DIS}_{A,B,C}\}, \{\mathsf{K}^0_{A,B}\}, \{\mathsf{K}^1_{A,B}\} >$ be a frame. An assignment in $\mathcal{F}$ is a function f defined on VAR such that $f(x^A) \in \mathsf{D}^A$.

The set of all assignments in a given frame is denoted by ASG. If $y \in VAR$, then f^y_a is defined by $f^y_a(x) = x$, if $x \neq y$, $f^y_a(y) = a$.

Definition 8.7 Let $\mathcal{F} = < \{\mathsf{D}^A\}, \{\mathsf{AP}_{A,B}\}, \{\mathsf{PRO}^0_{A,B}\}, \{\mathsf{PRO}^0_{A,B}\}, \{\mathsf{PAIR}_{A,B}\}, \{\mathsf{DIS}_{A,B,C}\}, \{\mathsf{K}^0_{A,B}\}, \{\mathsf{K}^1_{A,B}\} >$ be a frame. $< \mathcal{F}, VAL >$ is said to be a type structure model (or just a model) based on $\mathcal{F}$ iff VAL is the valuation function from $TERM \times ASG$ to $\bigcup_A \mathsf{D}^A$ such that

1. $VAL(x, f) = f(x)$;

2. $\mathsf{AP}_{A,B}(VAL((\lambda x M), f), a) = VAL(M, f^x_a)$, $\forall a \in \mathsf{D}^A$;

3. $VAL((M^{(A\supset B)}, N^B), f) = \mathsf{AP}_{A,B}(VAL(M, f), VAL(N, f))$;

4. $VAL(< M^A, N^B >, f) = \mathsf{PAIR}_{A,B}(VAL(M, f), VAL(N, f))$;

5. $VAL((M^{(A\wedge B)})_i, f) = \mathsf{PRO}^i_{A,B}(VAL(M, f))$, $i = 0, 1$;

6. $VAL(K^0_{A,B}(M), f) = \mathsf{K}^0_{A,B}(VAL(M, f))$;

7. $VAL(K^1_{A,B}(M), f) = \mathsf{K}^1_{A,B}(VAL(M, f))$;

8. $VAL([M^{(A\supset C)}, N^{(B\supset C)}], f) = \mathsf{DIS}_{A,B,C}(VAL(M, f), VAL(N, f))$;

9. $VAL(M^{\sim\sim A}, f) = VAL(M^A, f)$;

10. $VAL(M^{\sim(A\wedge B)}, f) = VAL(M^{\sim A\vee\sim B}, f)$;

11. $VAL(M^{\sim(A\vee B)}, f) = VAL(M^{\sim A\wedge\sim B}, f)$;

12. $VAL(M^{\sim(A\supset B)}, f) = VAL(M^{A\wedge\sim B}, f)$;

13. if $x, y \notin FV(M)$, then
$\mathsf{AP}(\mathsf{DIS}(VAL((\lambda x^A(M^{(A\vee B)\supset C}, K^0_{A,B}(x))), f), VAL((\lambda y^B(M, K^1_{A,B}(y))), f)), a) =$
$\mathsf{AP}(VAL(M, f), a)$, $\forall a \in \mathsf{D}^{A\vee B}$;

14. $\mathsf{AP}(\mathsf{DIS}(VAL((\lambda x^A K^0_{A,B}(x)), f), VAL((\lambda y^B K^1_{A,B}(y)), f)), a) = a$, $\forall a \in \mathsf{D}^{A\vee B}$.

Lemma 8.8 (i) $VAL(M[x := N], f) = VAL(M, f^x_{VAL(N,f)})$, if $BV(M) \cap FV(N) = \emptyset$.
(ii) $VAL(M[x := y], f^y_a) = VAL(M, f^x_a)$, if $y \notin BV(M) \cap FV(M)$.

PROOF (i) By a straightforward induction on the generation of M, for fixed N. Here, by way of example, we consider just three cases:

- $M = K^0_{A,B}(G)$:
$VAL(K^0_{A,B}(G)[x := N], f) = VAL(K^0_{A,B}(G[x := N]), f)$
$= \mathsf{K}^0_{A,B}(VAL(G[x := N], f))$
$= \mathsf{K}^0_{A,B}(VAL(G, f^x_{VAL(N,f)}))$, by the induction hypothesis
$= VAL(K^0_{A,B}(G), f^x_{VAL(N,f)})$.

- $M = [G_1, G_2]$:
$VAL([G_1, G_2][x := N], f) = VAL([G_1[x := N], G_2[x := N]], f)$
$= \mathsf{DIS}(VAL(G_1[x := N], f), VAL(G_2[x := N], f))$
$= \mathsf{DIS}(VAL(G_1, f^x_{VAL(N,f)}), VAL(G_2, f^x_{VAL(N,f)}))$, by the induction hypothesis
$= VAL([G_1, G_2], f^x_{VAL(N,f)})$.

- $M = \lambda y G$, $y \notin FV(N)$: By induction on the generation of G. E.g.
$G = \lambda z G_1$:
$\mathsf{AP}(VAL((\lambda y(\lambda z G_1))[x := N], f), a) =$
$\mathsf{AP}(VAL(\lambda y(\lambda z G_1)[x := N], f), a) =$
$\mathsf{AP}(VAL(\lambda y(\lambda z G_1), f^x_{VAL(N,f)}), a)$, since by the induction hypothesis,
$VAL(\lambda z G_1[x := N], f) = VAL(\lambda z G_1, f^x_{VAL(N,f)})$.

(ii) By (i). □

Definition 8.9 Let $\mathcal{M} =< \mathcal{F}, VAL >$ be a model. We say that $M = N$ holds in $\mathcal{M}$ under assignment f ($\mathcal{M} \models M = N[f]$) iff $VAL(M, f) = VAL(N, f)$. $M = N$ is said to be valid in $\mathcal{M}$ ($\mathcal{M} \models M = N$) iff $\mathcal{M} \models M = N[f]$, for all $f \in ASG$. $M = N$ is said to be valid in a certain class of models, if $\mathcal{M} \models M = N$, for each $\mathcal{M}$ in this class.

8.1.3 Intended models for λ^c

Now, the idea is not just to characterize provable equality of terms in λ^c by validity in all models based on type structure frames, but rather to prove a characterization theorem wrt models based on particular frames, viz. 'full constructive type structures over infinite sets': $\mathcal{F}_c =< \{\mathbf{D}^A\}, \{\mathbf{AP}_{A,B}\}, \{\mathbf{PRO}^0_{A,B}\}, \{\mathbf{PRO}^1_{A,B}\}, \{\mathbf{PAIR}_{A,B}\}, \{\mathbf{DIS}_{A,B,C}\}, \{\mathbf{K}^0_{A,B}\}, \{\mathbf{K}^1_{A,B}\} >$, where

- for every $p \in PROP \cup \{\perp\}$, $\mathbf{D}^p$, $\mathbf{D}^{\sim p}$ are infinite sets;
- $\mathbf{D}^{A \wedge B} = \mathbf{D}^A \times \mathbf{D}^B$;
- $\mathbf{D}^{A \vee B} = \mathbf{D}^A \dot{\cup} \mathbf{D}^B$, if $\mathbf{D}^A \neq \mathbf{D}^B$, otherwise $\mathbf{D}^{A \vee B} = \mathbf{D}^A$;
- $\mathbf{D}^{A \supset B} = (\mathbf{D}^B)^{\mathbf{D}^A}$;
- $\mathbf{D}^{\sim\sim A} = \mathbf{D}^A$;
- $\mathbf{D}^{\sim(A \wedge B)} = \mathbf{D}^{\sim A \vee \sim B}$;
- $\mathbf{D}^{\sim(A \vee B)} = \mathbf{D}^{\sim A \wedge \sim B}$;
- $\mathbf{D}^{\sim(A \supset B)} = \mathbf{D}^{A \wedge \sim B}$;
- $\mathbf{AP}_{A,B}(a, b) = a(b)$;
- $\mathbf{PRO}^0_{A,B}(< a, b >) = a$;
- $\mathbf{PRO}^1_{A,B}(< a, b >) = b$;
- $\mathbf{PAIR}_{A,B}(a, b) =< a, b >$;
- $\mathbf{DIS}_{A,B,C}(a, b) = a \restriction (\mathbf{D}^A - \mathbf{D}^B) \cup b \restriction (\mathbf{D}^B - \mathbf{D}^A)$, if $\mathbf{D}^A \neq \mathbf{D}^B$, otherwise $\mathbf{DIS}_{A,B,C}(a, b) = a$ or $\mathbf{DIS}_{A,B,C}(a, b) = b$;
- $\mathbf{K}^0_{A,B}(a) = b$ chosen from $\mathbf{D}^A - \mathbf{D}^B$, if $a \in \mathbf{D}^A \cap \mathbf{D}^B$ and $\mathbf{D}^A \neq \mathbf{D}^B$, otherwise $\mathbf{K}^0_{A,B}(a) = a$;
- $\mathbf{K}^1_{A,B}(a) = b$ chosen from $\mathbf{D}^B - \mathbf{D}^A$, if $a \in \mathbf{D}^A \cap \mathbf{D}^B$ and $\mathbf{D}^A \neq \mathbf{D}^B$, otherwise $\mathbf{K}^1_{A,B}(a) = a$.

Here $\dot{\cup}$ denotes disjoint set-union (i.e. $\Gamma_1 \dot{\cup} \Gamma_2 = (\Gamma_1 - \Gamma_2) \cup (\Gamma_2 - \Gamma_1)$). It can easily be verified that $\mathcal{F}_c$ in fact is a type structure frame. Observe that $a \in \mathbf{D}^{A \vee B}$ iff ($a = \mathbf{K}^0_{A,B}(a)$ or $a = \mathbf{K}^1_{A,B}(a)$). Let $\vdash_{\lambda^c}$ denote the provability relation induced by λ^c's axiom schemes and rules.

Theorem 8.10 (soundness) If $\vdash_{\lambda^c} M = N$, then $M = N$ is valid in the class of all models.

PROOF By induction on the length of proofs in λ^c. We must show that every axiom is valid in every model, and that the rules of inference preserve validity. We shall consider the non-obvious cases which are not already covered by [Friedman 1975]. Let $\mathcal{M} = < \{\mathsf{D}^A\}, \{\mathsf{AP}_{A,B}\}, \{\mathsf{PRO}^0_{A,B}\}, \{\mathsf{PRO}^1_{A,B}\}, \{\mathsf{PAIR}_{A,B}\}, \{\mathsf{DIS}_{A,B,C}\}, \{\mathsf{K}^0_{A,B}\}, \{\mathsf{K}^1_{A,B}\}, VAL >$ be a model.

- $< sur >$: $VAL((< M,N >)_0, (< M,N >)_1 >, f)$
  $= \mathsf{PAIR}(VAL((< M,N)_0, f), VAL((< M,N >)_1, f))$
  $= \mathsf{PAIR}(\mathsf{PRO}^0(\mathsf{PAIR}(VAL(M,f), VAL(N,f)))),$
  $\mathsf{PRO}^1(\mathsf{PAIR}(VAL(M,f), VAL(N,f)))$
  $= \mathsf{PAIR}(VAL(M,f), VAL(N,f)) = VAL(< M,N >, f)$.
- (0): $VAL(([M^{A\supset C}, N^{B\supset C}], K^0_{A,B}(G)), f)$
  $= \mathsf{AP}(\mathsf{DIS}(VAL(M,f), VAL(N,f)), VAL(K^0(G), f))$
  $= \mathsf{AP}(\mathsf{DIS}(VAL(M,f), VAL(N,f)), \mathsf{K}^0(VAL(G,f)))$
  $= \mathsf{AP}(VAL(M,f), VAL(G,f))$, by (red $\vee$) (i)
  $= VAL((M,G), f)$.
- (1): Analogous to the previous case, using (red $\vee$) (ii).
- (ϑ_1): Suppose that $x, y \notin FV(M)$. Then
  $VAL(([(\lambda x^A(M^{A\vee B)\supset C}, K^0_{A,B}(x))), (\lambda y^B(M, K^1_{A,B}(y)))], G), f)$
  $= \mathsf{AP}(\mathsf{DIS}(VAL((\lambda x(M, K^0(x))), f), VAL((\lambda y(M, K^1(y))), f)), VAL(G,f))$
  $= \mathsf{AP}(VAL(M,f), VAL(G,f)) = VAL((M,G), f)$.
- (ϑ_2): Use clause 14 of Definition 8.7. □

8.1.4 The canonical model and completeness

Before we proceed to extending Friedman's completeness proof for $\lambda_\supset$ to λ^c, here is a brief description of the structure of the proof. We first define a certain canonical type structure model $\mathcal{M}_0$ for λ^c which is based on a frame $\mathcal{F}_0$, and show that λ^c is complete wrt this model. Next, it is shown that 'partial homomorphisms' preserve validity in type structure models, and eventually it is proved that there exists a partial homomorphism from models based on frames $\mathcal{F}_c$ onto $\mathcal{M}_0$. Let $\mid M \mid = \{N \mid\vdash_{\lambda^c} M = N\}$. By the rules 1, $\mid M \mid$ is the equivalence class of M wrt provable equality with M in λ^c.

Definition 8.11 $\mathcal{F}_0 = < \{\mathsf{D}^A\}, \{\mathsf{AP}_{A,B}\}, \{\mathsf{PRO}^0_{A,B}\}, \{\mathsf{PRO}^1_{A,B}\}, \{\mathsf{PAIR}_{A,B}\}, \{\mathsf{DIS}_{A,B,C}\}, \{\mathsf{K}^0_{A,B}\}, \{\mathsf{K}^1_{A,B}\} >$ is defined as follows:

- $\mathsf{D}^A = \{\mid M \mid \mid M$ is of type $A\}$, if there is no $N^{B_1} \in \mid M \mid$ such that B_1 is of the form $(B \wedge C)$ or $(B \vee C)$;
- $\mathsf{D}^{(A\wedge B)} = \{\mid M \mid \mid M$ is of type $A \wedge B$ and there exists $< N_1, N_2 >$ such that $\vdash_{\lambda^c} M = < N_1, N_2 >\}$;

- $\mathsf{D}^{(A\vee B)} = \{| M | \mid M$ is of type $A \vee B$ and there exists $K^0_{A,B}(N_1)$ or $K^1_{A,B}(N_2)$ such that $\vdash_{\lambda^c} M = K^0_{A,B}(N_1)$ resp. $\vdash_{\lambda^c} M = K^1_{A,B}(N_2)\}$;
- $\mathsf{AP}_{A,B}(| M^{A\supset B} |, | N^A |) = | (M, N) |$;
- $\mathsf{PRO}^0_{A,B}(| M^{A\wedge B} |) = | (M)_0 |$;
- $\mathsf{PRO}^1_{A,B}(| M^{A\wedge B} |) = | (M)_1 |$;
- $\mathsf{PAIR}_{A,B}(| M^A |, | N^B |) = |< M, N >|$;
- $\mathsf{DIS}_{A,B,C}(| M^{A\supset C} |, | N^{A\supset C} |) = | [M, N] |$;
- $\mathsf{K}^0_{A,B}(| M^A |) = | K^0(M) |$;
- $\mathsf{K}^1_{A,B}(| M^B |) = | K^1(M) |$.

Lemma 8.12 $\mathcal{F}_0$ is a frame.

PROOF Clearly, D^A is a non-empty set, and $\mathsf{AP}_{A,B}$, $\mathsf{PAIR}_{A,B}$, $\mathsf{DIS}_{A,B,C}$ are functions with appropriate domain and range, for all types A, B, C. For extensionality see the proof of Lemma 7 in [Friedman 1975]. (red $\wedge$) (i): Use $< 0 >$. (red $\wedge$) (ii): Use $< 1 >$. (red $\vee$) (i): Suppose that $| M | \in \mathsf{D}^{A\supset C}$, $| N | \in \mathsf{D}^{B\supset C}$, and $| G | \in \mathsf{D}^A$. We must show that $| ([M, N], K^0_{A,B}(G)) | = | (M, N) |$. Assume that $\vdash_{\lambda^c} M = M_1$, $\vdash_{\lambda^c} N = N_1$, and $\vdash_{\lambda^c} G = G_1$. Then, by (0) and the rules 2, $([M_1, N_1], K^0_{A,B}(G_1)) \in | ([M, N], K^0_{A,B}(G)) |$ implies $(M_1, G_1) \in | (M, G) |$, and $(M_1, G_1) \in | (M, G) |$ implies $([M_1, N_1], K^0_{A,B}(G_1)) \in | (M, G) |$. (red $\vee$) (ii): analogous. Finally, suppose $\mathsf{D}^A = \mathsf{D}^B$. Then $M^A = M^B$ and we have e.g. $\mathsf{K}^0_{A,B}(| M |) = | K^0_{A,A}(M) | = | K^1_{A,B}(M) | = \mathsf{K}^1_{A,B}(| M |)$. □

Definition 8.13 A function $g : VAR \longrightarrow TERM$ is called a substitution, if $g(x)$ and x are of the same type. A substitution is called regular, if for pairwise distinct variables x, y, $FV(g(x)) \cap FV(g(y)) = \emptyset$.

Let $M(g)$ denote the result of simultaneously replacing every free occurrence of each variable in x by $g(x)$. By (α) and the rules 1, we have

Lemma 8.14 If $M \in TERM$ and Γ is a finite set of variables, then there is an N such that $\vdash_{\lambda^c} M = N$, $FV(M) = FV(N)$, and $BV(N) \cap \Gamma = \emptyset$.

Definition 8.15 Let f be an assignment in $\mathcal{F}_0$, and let, by the previous lemma, g be a regular substitution such that $f(x) = | g(x) |$, for every $x \in VAR$. For a given term M, choose a term N such that $\vdash_{\lambda^c} M = N$ and for every $x \in FV(N)$, $BV(N) \cap FV(g(x)) = \emptyset$. Then $VAL(M, f)$ is defined by $VAL(M, f) = | N(g) |$.

As in [Friedman 1975, p. 26] it can be shown that $VAL : TERM \times ASG \longrightarrow \bigcup_A \mathsf{D}^A$, and $\vdash_{\lambda^c} M = N$ implies $VAL(M, f) = VAL(N, f)$.

Lemma 8.16 $\mathcal{M}_0 =< \mathcal{F}_0, VAL >$ is a type structure model.

PROOF We must check the conditions on VAL; we consider those conditions not already dealt with in [Friedman 1975]. Let g be a regular substitution and $f(x) = \mid g(x) \mid$, for $f \in ASG$. Choose x_1, y_1, M_1, N_1 such that $\vdash_{\lambda^c} x = x_1$, $\vdash_{\lambda^c} y = y_1$, $\vdash_{\lambda^c} M = M_1$, $\vdash_{\lambda^c} N = N_1$, x_1, $y_1 \notin FV(M_1)$, and $BV(M_1) \cap FV(g(x)) = BV(N_1) \cap FV(g(x)) = \emptyset$, for every $x \in FV(M_1) \cup FV(N_1)$.

4: $VAL(< M, N >, f) = \mid < M_1, N_1 > (g) \mid =$
$\mathsf{PAIR}(\mid M_1(g) \mid, \mid N_1(g) \mid) = \mathsf{PAIR}(VAL(M, f), VAL(N, f))$.

5: $VAL((M)_i, f) = \mid (M_1)_i(g) \mid = \mathsf{PRO}^i(\mid M_1(g) \mid) = \mathsf{PRO}^i(VAL(M, f))$.

6: $VAL(K^0_{A,B}(M), f) = \mid K^0(M_1)(g) \mid =$
$\mathsf{K}^0(\mid M_1(g) \mid) = \mathsf{K}^0(VAL(M, f))$.

7: Analogous to the previous case.

8: $VAL([M, N], f) = \mid [M_1, N_1] \mid = \mid [M_1(g), N_1(g)] \mid =$
$\mathsf{DIS}(\mid M_1(g) \mid, \mid N_1(g) \mid) = \mathsf{DIS}(VAL(M, f), VAL(N, f))$.

9 – 12: Use $(\sim)$.

13: Suppose that x, $y \notin FV(M)$.
$\mathsf{AP}(\mathsf{DIS}(VAL((\lambda x^A(M^{(A \vee B) \supset C}, K^0_{A,B}(x))), f), VAL((\lambda y^B(M, K^1_{A,B}(y))), f), \mid N \mid)$
$= \mid ([(\lambda x_1(M_1 K^0(x_1)))(g), (\lambda y_1(M_1 K^1(y_1)))(g)], N) \mid$
$= \mid (M_1(g), N) \mid$ by (ϑ_1)
$= \mathsf{AP}(\mid M_1(g) \mid, \mid N \mid) = \mathsf{AP}(VAL(M, f), \mid N \mid)$.

14: $\mathsf{AP}(\mathsf{DIS}(VAL((\lambda x^A K^0_{A,B}(x_1)), f), VAL((\lambda y^B K^1_{A,B}(y)), f)), \mid N \mid)$
$= \mid ([(\lambda x_1 K^0(x_1))(g), (\lambda y_1 K^1(y_1))(g)], N) \mid$
$= \mid N \mid$ by (ϑ_2). □

Theorem 8.17 (completeness) If $M = N$ is valid in the class of all models, then $\vdash_{\lambda^c} M = N$.

PROOF Suppose $\not\vdash_{\lambda^c} M = N$. Choose M_1, N_1 such that $\vdash_{\lambda^c} M = N_1$, $\vdash_{\lambda^c} N = N_1$, and $BV(M_1) \cap FV(M_1) = BV(N_1) \cap FV(N_1) = \emptyset$. Then $VAL(M, f) = \mid M_1 \mid \neq \mid N_1 \mid = VAL(N, f)$, for $f(x) = \mid \mathsf{id}(x) \mid$, for all $x \in VAR$, where id is the identity function on VAR. Thus, $\mathcal{M}_0 \not\models M = N$. □

8.1.5 Completeness wrt intended models

Definition 8.18 Let $\mathcal{F} = < \{\mathsf{D}^A\}, \{\mathsf{AP}_{A,B}\}, \{\mathsf{PRO}^0_{A,B}\}, \{\mathsf{PRO}^1_{A,B}\}, \{\mathsf{PAIR}_{A,B}\}, \{\mathsf{DIS}_{A,B,C}\}, \{\mathsf{K}^0_{A,B}\}, \{\mathsf{K}^1_{A,B}\} >$ and $\mathcal{F}^* = < \{\mathsf{D}^{*A}\}, \{\mathsf{AP}^*_{A,B}\}, \{\mathsf{PRO}^{*0}_{A,B}\}, \{\mathsf{PRO}^{*1}_{A,B}\}, \{\mathsf{PAIR}^*_{A,B}\}, \{\mathsf{DIS}^*_{A,B,C}\}, \{\mathsf{K}^{*0}_{A,B}\}, \{\mathsf{K}^{*1}_{A,B}\} >$ be frames; let $\mathcal{M} = < \mathcal{F}, VAL >$ and $\mathcal{M}^* = < \mathcal{F}^*, VAL^* >$ be models. A family $\{f_A\}$ is called a partial homomorphism from $\mathcal{M}$ onto $\mathcal{M}^*$ iff

- for each type A, f_A is a partial function from D^A onto D^{*A};

- if $f_{A\supset B}(a)$ exists, then $f_B(\mathsf{AP}_{A,B}(a,b)) = \mathsf{AP}^*_{A,B}(f_{A\supset B}(a), f_A(b))$, for all b in the domain of f_A,
- if f_A, f_B exist, then $f_{A\wedge B}(\mathsf{PAIR}_{A,B}(a,b)) = \mathsf{PAIR}^*_{A,B}(f_A(a), f_B(b))$;
- if $f_{A\wedge B}(a)$ exists, then $f_A(\mathsf{PRO}^0_{A,B}(a)) = \mathsf{PRO}^{*0}_{A,B}(f_{A\wedge B}(a))$;
- if $f_{A\wedge B}(a)$ exists, then $f_B(\mathsf{PRO}^1_{A,B}(a)) = \mathsf{PRO}^{*1}_{A,B}(f_{A\wedge B}(a))$;
- if $f_{A\supset C}(a)$, $f_{B\supset C}(b)$ exist, then $f_{(A\vee B)\supset C}(\mathsf{DIS}_{A,B,C}(a,b))$ $= \mathsf{DIS}^*_{A,B,C}(f_{A\supset C}(a), f_{B\supset C}(b))$;
- if $f_A(a)$ exists, then $f_{A\vee B}(K^0_{A,B}(a)) = \mathsf{K}^{*0}_{A,B}(f_A(a))$; and
- if $f_B(b)$ exists, then $f_{A\vee B}(K^1_{A,B}(b)) = \mathsf{K}^{*1}_{A,B}(f_B(b))$.

Lemma 8.19 Let $\mathcal{M}$, $\mathcal{M}^*$ be as in the previous definition, and let $\{f_A\}$ be a partial homomorphism from $\mathcal{M}$ onto $\mathcal{M}^*$. If g resp. $g*$ is an assignment in $\mathcal{F}$, resp. $\mathcal{F}^*$, and $f_A(g(x^A)) = g^*(x)$, then $f_A(VAL(M^A, g)) = VAL^*(M, g^*)$.

PROOF By induction on the generation of M. Again, we restrict ourselves to the cases not already covered by [Friedman 1975]. Note that we may assume $f_A(g(x^A)) = g^*(x)$, because f_A is onto.

- $M = \langle N^A, G^B\rangle$: $f_{A\wedge B}(VAL(\langle N,G\rangle, g))$
  $= f_{A\wedge B}(\mathsf{PAIR}(VAL(N,g), VAL(G,g)))$
  $= \mathsf{PAIR}^*_{A,B}(f_A(VAL(N,g)), f_B(VAL(G,g)))$
  $= \mathsf{PAIR}^*_{A,B}(VAL^*(N,g^*), VAL^*(G,g^*))$ by the induction hypothesis
  $= VAL^*(\langle N,G\rangle, g^*)$.
- $M = (M^{A\wedge B})^A_0$: $f_A(VAL((M)_0, g)) = f_A(\mathsf{PRO}^0(VAL(M,g)))$
  $= \mathsf{PRO}^{*0}(f_{A,B}(VAL(M,g)))$
  $= \mathsf{PRO}^{*0}(VAL^*(M,g^*))$ by the induction hypothesis
  $= VAL^*((M)_0, g^*)$.
- $M = (M^{A\wedge B})^A_1$: Analogous to the previous case.
- $M = K^0_{A,B}(N)$: $f_{A\vee B}(VAL(K^0(N), g))$
  $= f_{A\vee B}(\mathsf{K}^0_{A,B}(VAL(N,g)))$
  $= \mathsf{K}^{*0}_{A,B}(f_A(VAL(N,g)))$
  $= \mathsf{K}^{*0}_{A,B}(VAL^*(N,g^*))$ by the induction hypothesis
  $= VAL^*(K^0_{A,B}(N), g^*)$.
- $M = K^1_{A,B}(N)$: Analogous to the previous case.
- $M = [N,G]^{(A\vee B)\supset C}$: $f_{(A\vee B)\supset C}(VAL([N,G], g))$
  $= f(\mathsf{DIS}(VAL(N,g), VAL(G,g)))$
  $= \mathsf{DIS}^*(f_{A\supset C}(VAL(N,g)), f_{B\supset C}(VAL(G,g)))$
  $= \mathsf{DIS}^*(VAL^*(N,g^*), VAL^*(G,g^*))$ by the induction hypothesis
  $= VAL^*([N,G], g^*)$. □

Lemma 8.20 Let $\mathcal{M}_1$, $\mathcal{M}_2$ be models. If there exists a partial homomorphism from $\mathcal{M}_1$ onto $\mathcal{M}_2$, then $\mathcal{M}_1 \models M = N$ implies $\mathcal{M}_2 \models M = N$.

PROOF By the previous lemma, cf. [Friedman 1975, p. 28]. □

Theorem 8.21 Let $\mathcal{M}$ be a type structure model based on $\mathcal{F}_c$. Then $\vdash_{\lambda^c} M = N$ iff $\mathcal{M} \models M = N$.

PROOF In view of the previous two theorems it is enough to show that $\mathcal{M} \models M = N$ implies $\mathcal{M}_0 \models M = N$. We show this by defining a partial homomorphism $\{f_A\}$ from $\mathcal{M}$ onto $\mathcal{M}_0$; $\{f_A\}$ is defined by induction on the type A.

- $A = p$, $A =\sim p$ ($p \in PROP \cup \{\bot\}$): f_A is any function from $\mathbf{D}^A$ onto D^A. Such a function exists, since $\mathbf{D}^A$ is infinite and D^A is denumerable.
- $A = (B \wedge C)$: If $f_B(b)$, $f_C(c)$ exist, then $f_{B \wedge C}(< b, c >) = f_{B\wedge C}(\mathbf{PAIR}(b, c))$ is defined as $\mathsf{PAIR}_{B,C}(f_B(c), f_C(c))$
- $A = (B \vee C)$: Recall that $a \in \mathbf{D}^{A\vee B}$ iff ($a = \mathbf{K}^0_{A,B}(a)$ or $a = \mathbf{K}^1_{A,B}(a)$). Define $f_{B\vee C}(\mathbf{K}^0_{B,C}(a)) = \mathsf{K}^0_{B,C}(f_B(a))$, if $f_B(a)$ exists; $f_{B\vee C}(\mathbf{K}^1_{B,C}(a)) = \mathsf{K}^1_{B,C}(f_C(a))$, if $f_C(a)$ exists.
- $A = (B \supset C)$: Suppose that f_B, f_C have already been defined. Then $f_{B\supset C}(a)$ is defined as the unique member of $\mathsf{D}^{(B\supset C)}$ (if it exists) such that $f_C(a(b)) = \mathsf{AP}_{B,C}(f_{B\supset C}(a), f_B(b))$, for all b in the domain of f_B.
  If B is of the form $B_1 \vee B_2$, we require in addition that, if $f_{B_1\supset C}(a)$, $f_{B_2\supset C}(b)$ exist, then $f_{(B_1\vee B_2)\supset C}(\mathbf{DIS}_{B_1,B_2,C}(a, b)) =$
  $\mathsf{DIS}_{B_1,B_2,C}(f_{B_1\supset C}(a), f_{B_2\supset C}(b))$.
- $A =\sim\sim B$: $f_{\sim\sim B}(x) = f_B(x)$.
- $A =\sim (B \wedge C)$: $f_{\sim(B\wedge C)}(x) = f_{\sim B\vee\sim C}(x)$.
- $A =\sim (B \vee C)$: $f_{\sim(B\vee C)}(x) = f_{\sim B\wedge\sim C}(x)$.
- $A =\sim (B \supset C)$: $f_{\sim(B\supset C)}(x) = f_{B\wedge\sim C}(x)$.

From this definition together with the following equations it follows immediately that $\{f_A\}$ is a partial homomorphism:

$$
\begin{aligned}
& f_A(\mathbf{PRO}^0_{A,B}(< a, b >)) \\
&= f_A(a) \\
&= \mathsf{PRO}^0_{A,B}(\mathsf{PAIR}_{A,B}(f_A(a), f_B(b))) \\
&= \mathsf{PRO}^0_{A,B}(f_{A\wedge B}(\mathbf{PAIR}_{A,B}(a, b))) \\
&= \mathsf{PRO}^0_{A,B}(f_{A\wedge B}(< a, b >));
\end{aligned}
$$

$$
\begin{aligned}
& f_B(\mathbf{PRO}^1_{A,B}(< a, b >)) \\
&= f_B(b) \\
&= \mathsf{PRO}^1_{A,B}(\mathsf{PAIR}_{A,B}(f_A(a), f_B(b)))
\end{aligned}
$$

$= \mathsf{PRO}^1_{A,B}(f_{A\wedge B}(\mathbf{PAIR}_{A,B}(a,b)))$
$= \mathsf{PRO}^1_{A,B}(f_{A\wedge B}(< a, b >))$.

It remains to be shown that f_A is onto, for every type A. This follows from the definition of f_p, $f_{\sim p}$ ($p \in PROP \cup \{\bot\}$) and the definition of $\mathcal{F}_0$, except for the case that $A = (B \supset C)$. Thus, assume $d \in \mathsf{D}^{(B\supset C)}$. Choose $a \in \mathbf{D}^{(B\supset C)}$ such that for every b in the domain of f_B, $a(b) \in f_C^{-1}(\mathsf{Ap}(d, f_B(b)))$. Then $f_{(B\supset C)}(a) = d$. □

As in [Friedman 1975, Theorem 4] for the case of $\lambda_\supset$ and 'full type structures over finite sets', it can be shown that equality in λ^c is not characterized by 'full constructive type structures over finite sets'.

8.2 Formulas-as-types for N$^-$

We show that every $M \in TERM$ encodes a proof in Nelson's propositional logic $\mathbf{N}^-$, and vice versa. We do this for the sequent calculus from Chapter 2, in the propositional language $\{\sim, \wedge, \vee, \supset, \bot\}$. For the reader's convenience, we present this calculus here once more:

(*id*); (*cut*);

$<\rightarrow\wedge>$	$X \rightarrow A \quad Y \rightarrow B \vdash XY \rightarrow (A\wedge B)$;
$<\wedge\rightarrow>$	$XA \rightarrow C \vdash X(A\wedge B) \rightarrow C$,
	$XB \rightarrow C \vdash X(A\wedge B) \rightarrow C$;
$<\rightarrow\vee>$	$X \rightarrow A \vdash X \rightarrow (A\vee B)$,
	$X \rightarrow B \vdash X \rightarrow (A\vee B)$;
$<\vee\rightarrow>$	$XA \rightarrow C \quad YB \rightarrow C \vdash XY(A\vee B) \rightarrow C$;
$<\rightarrow\supset>$	$XA \rightarrow B \vdash X \rightarrow (A\supset B)$;
$<\supset\rightarrow>$	$Y \rightarrow A \quad XB \rightarrow C \vdash X(A\supset B)Y \rightarrow C$;
$<\rightarrow\sim\sim>$	$X \rightarrow A \vdash X \rightarrow\sim\sim A$;
$<\sim\sim\rightarrow>$	$XA \rightarrow B \vdash X \sim\sim A \rightarrow B$;
$<\rightarrow\sim\wedge>$	$X \rightarrow\sim A \vdash X \rightarrow\sim (A\wedge B)$,
	$X \rightarrow\sim B \vdash X \rightarrow\sim (A\wedge B)$;
$<\sim\wedge\rightarrow>$	$X \sim A \rightarrow C \quad Y \sim B \rightarrow C \vdash XY \sim (A\wedge B) \rightarrow C$;
$<\rightarrow\sim\vee>$	$X \rightarrow\sim A \quad Y \rightarrow\sim B \vdash XY \rightarrow\sim (A\vee B)$;
$<\sim\vee\rightarrow>$	$X \sim A \rightarrow C \vdash X \sim (A\vee B) \rightarrow C$,
	$X \sim B \rightarrow C \vdash X \sim (A\vee B) \rightarrow C$;
$<\rightarrow\sim\supset>$	$X \rightarrow A \quad Y \rightarrow\sim B \vdash XY \rightarrow\sim (A\supset B)$;
$<\sim\supset\rightarrow>$	$XA \rightarrow C \vdash X \sim (A\supset B) \rightarrow C$;
	$X \sim B \rightarrow C \vdash X \sim (A\supset B) \rightarrow C$;

P; **C**; **M**.

Definition 8.22 M^A is a construction of a sequent $A_1 \ldots A_n \rightarrow A$ iff there are at most free variable occurrences $x_1^{A_1}, \ldots, x_n^{A_n} \in FV(M)$.

Let $PROOF_{\mathbf{N}^-}$ denote the set of sequent proofs in $\mathbf{N}^-$.

Theorem 8.23 Given a proof in $PROOF_{\mathbf{N}^-}$ of $s = A_1 \dots A_n \to A$, one can find a construction M of s, and conversely.

PROOF We shall inductively define encoding functions $f : PROOF_{\mathbf{N}^-} \longrightarrow TERM$, $g : TERM \longrightarrow PROOF_{\mathbf{N}^-}$ such that it can readily be seen that $f(\Pi)$ is a construction of Π, and $g(M)$ is a sequent of which M is a construction. We use Π, Π_1, Π_2 etc. to denote proofs. The function f is inductively defined by:

- $\Pi = \vdash A \to A$: $f(\Pi) = v_1^A$.
- $\Pi = \frac{\frac{\Pi_1}{Y\to A}\quad\frac{\Pi_2}{XAZ\to B}}{XYZ\to B}$: $f(\Pi) = f(\frac{\Pi_2}{XAZ\to B}),[v_i^A := f(\frac{\Pi_1}{Y\to A})]$,
  where v_i is the first variable of type A in $FV(f(\frac{\Pi_2}{XAZ\to B}))$, if there is such a variable; otherwise $v_i^A = v_1^A$.
- $\Pi = \frac{\frac{\Pi_1}{X\to A}\quad\frac{\Pi_2}{Y\to B}}{XY\to(A\wedge B)}$: $f(\Pi) =< f(\frac{\Pi_1}{X\to A}), f(\frac{\Pi_2}{Y\to B}) >$.
- $\Pi = \frac{\frac{\Pi_1}{XA\to C}}{X(A\wedge B)\to C}$: $f(\Pi) = f(\frac{\Pi_1}{X\to A})[v_i^A := (v_j^{(A\wedge B)})_0]$,
  where v_j is the first fresh variable of type $A \wedge B$, and v_i is the first variable of type A in $FV(f(\frac{\Pi_1}{XA\to C}))$, if there is such a variable; otherwise $v_i^A = v_1^A$.
  The second rule of $< \wedge \to>$ is treated analogously.
- $\Pi = \frac{\frac{\Pi_1}{X\to A}}{X\to(A\vee B)}$: $f(\Pi) = K^0_{A,B}(f(\frac{\Pi_1}{X\to A}))$.
  The second rule of $<\to \vee >$ is dealt with analogously, using $K^1_{A,B}$.
- $\Pi = \frac{\frac{\Pi_1}{XA\to C}\quad\frac{\Pi_2}{YB\to C}}{XY(A\vee B)\to C}$: $f(\Pi) = ([< \lambda v_i^A f(\frac{\Pi_1}{XA\to C}), \lambda v_j^B f(\frac{\Pi_2}{YB\to C}) >], v_k^{A\vee B})$,
  where v_k is the first fresh variable of type $A \vee B$; v_i resp. v_j is the first variable of type A resp. type B in $FV(f(\frac{\Pi_1}{XA\to C}))$ resp. $FV(f(\frac{\Pi_2}{YB\to C}))$, if there are such variables; otherwise $v_i^A = v_1^A$, $v_j^B = v_1^B$.
- $\Pi = \frac{\frac{\Pi_1}{XA\to B}}{X\to(A\supset B)}$: $f(\Pi) = (\lambda v_i^A f(\frac{\Pi_1}{XA\to B}))$,
  where v_i is the first variable of type A in $FV(f(\frac{\Pi_1}{XA\to B}))$, if there is such a variable; otherwise $v_i^A = v_1^A$.
- $\Pi = \frac{\frac{\Pi_1}{Y\to A}\quad\frac{\Pi_2}{XB\to C}}{X(A\supset B)Y\to C}$: $f(\Pi) = f(\frac{\Pi_2}{XB\to C}),[v_i^B := (v_j^{A\supset B}, f(\frac{\Pi_1}{Y\to A}))]$,
  where v_j is the first fresh variable of type $A \supset B$, and v_i is the first variable of type B in $FV(f(\frac{\Pi_2}{XB\to C}))$, if there is such a variable; otherwise $v_i^B = v_1^B$.
- $\Pi = \frac{\frac{\Pi_1}{XAAY\to B}}{XAY\to B}$: $f(\Pi) = f(\frac{\Pi_1}{XAAY\to B})[v_i^A := v_k^A][v_j^A := v_k^A]$,
  where v_k is the first fresh variable of type A, and v_i, v_j are the first distinct variables of type A in $FV(f(\frac{\Pi}{XAAY\to B}))$, if there are such variables; otherwise $v_i^A = v_j^A = v_1^A$.

- The last step in Π is an application of $<\to\sim\wedge>$, $<\sim\wedge\to>$, $<\to\sim\vee>$, $<\sim\vee\to>$, $<\to\sim\supset>$, $<\sim\supset\to>$, respectively: This case is treated in analogy to the case of $<\to\vee>$, $<\vee\to>$, $<\to\wedge>$, $<\wedge\to>$, $<\to\wedge>$, $<\wedge\to>$, respectively.

- All remaining cases: take as encoding term the term provided by the induction hypothesis.

The function g is inductively defined as follows:

- $M = x^A \in VAR$: $g(M) = \vdash A \to A$.

- $M = <N^A, G^B>$: $g(M) = \frac{g(N) \quad g(G)}{XY \to A\wedge B}$, where X resp. Y is the sequence of types of the free variable occurrences (fvos) in $g(N)$ resp. $g(G)$.

- $M = (\lambda x N)^{A\supset B}$: $g(M) = \frac{\frac{g(N)^B}{XA \to B}}{X \to A\supset B}$, where X is the sequence of types of the fvos in $g(N)$.

- $M = (N^{(A\wedge B)})_0$: $g(M) = \frac{g(N) \quad \frac{A\to A}{A\wedge B \to A}}{X \to A}$, where X is the sequence of types of the fvos in $g(N)$.

- $M = (N^{(A\wedge B)})_1$: Similar to the previous case.

- $M = (x^{A\supset B}, G^A)$: $g(M) = \frac{g(G) \quad B\to B}{(A\supset B)X \to B}$, where X is the sequence of types of the fvos in $g(G)$.

- $M = (N^{A\supset B}, G^A)$, $N \notin VAR$: $g(M) = \frac{g(G) \quad \frac{g(N) \quad \frac{A\to A \quad B\to B}{(A\supset B)A\to B}}{XA\to B}}{XY\to B}$, where X resp. Y is the sequence of types of the fvos in $g(N)$ resp. $g(G)$.

- $M = K^i_{A,B}(N)$, $i = 0,1$: $g(M) = \frac{g(N)}{X \to A\vee B}$, where X is the sequence of types of the fvos in $g(N)$.

- $M = [G,N]^{(A\vee B)\supset C}$: $g(M) = \frac{\frac{\frac{g(G) \quad \frac{A\to A \quad C\to C}{(A\supset C)A\to C}}{XA\to C} \quad \frac{g(N) \quad \frac{B\to B \quad C\to C}{(B\supset C)B\to C}}{YB\to C}}{XY(A\vee B)\to C}}{XY\to (A\vee B)\supset C}$, where X resp. Y is the sequence of types of the fvos in $g(G)$ resp. $g(N)$. □

We conclude this chapter by two simple

EXAMPLES (i) The sequent $s_1 = (\sim p_1 \supset p_3) \sim (p_1 \vee p_2) \to p_3$ has exactly two distinct (*cut*)-free proofs in $\mathbf{N}^-$ without using structural rules, viz.

$$\frac{\frac{\sim p_1 \to \sim p_1 \quad p_3 \to p_3}{(\sim p_1 \supset p_3) \sim p_1 \to p_3}}{s_1} \quad \text{and} \quad \frac{\frac{\sim p_1 \to \sim p_1}{\sim(p_1\vee p_2) \to \sim p_1} \quad p_3 \to p_3}{s_1},$$

which are encoded by $(v_1^{\sim p_1 \supset p_3}, v_1^{\sim p_1})$ resp. $(v_1^{\sim p_1 \supset p_3}, (v_1^{\sim(p_1\vee p_2)})_0^{\sim p_1})$.

(ii) Also the sequent $s_2 = (p_1 \lor p_2) \supset p_3\,(p_1 \lor p_2) \to p_3$ has exactly two distinct (cut)-free proofs in $\mathbf{N}^-$ without using structural rules, viz.

$$\frac{\dfrac{\dfrac{p_1 \to p_1}{p_1 \to (p_1 \lor p_2)} \quad p_3 \to p_3}{(p_1 \lor p_2) \supset p_3\, p_1 \to p_3} \qquad \dfrac{\dfrac{p_2 \to p_2}{p_2 \to (p_1 \lor p_2)} \quad p_3 \to p_3}{(p_1 \lor p_2) \supset p_3\, p_2 \to p_3}}{(p_1 \lor p_2) \supset p_3\,(p_1 \lor p_2) \to p_3} \quad \text{and} \quad \frac{(p_1 \lor p_2) \to (p_1 \lor p_2) \quad p_3 \to p_3}{(p_1 \lor p_2) \supset p_3\,(p_1 \lor p_2) \to p_3},$$

which are encoded by

$M = ([< \lambda v_1^{p_1}(v_1^{(p_1 \lor p_2) \supset p_3} K^{0}_{p_1,p_2}(v_1^{p_1})), \lambda v_1^{p_2}(v_1^{(p_1 \lor p_2) \supset p_3} K^{1}_{p_1,p_2}(v_1^{p_2})) >], v_1^{(p_1 \lor p_2)})$, resp.
$N = (v_1^{(p_1 \lor p_2) \supset p_3}, v_1^{p_1 \lor p_2})$. Note that $M \gg N$.

Chapter 9

Monoid models and the informational interpretation of substructural propositional logics

After having motivated, presented, and discussed in the previous chapters various families of substructural propositional logics, we shall now take up again the idea of an ***informational interpretation*** of propositional logics in models based on abstract information structures. The abstract information structures to be considered in the present chapter are semilattice-ordered groupoids $< I, \cdot, \cap, 1 >$ as introduced in [Došen 1989] which are semilattice-ordered monoids. Such semilattice-ordered monoids will be called ***slomos***, and models based on ***slomos*** will be called ***monoid models***. An informational interpretation of all the substructural logical systems we have dealt with will be developed, based on a suitable understanding of ***slomos***. We shall characterize all these systems by appropriate classes of monoid models. In the intuitionistic minimal and the intuitionistic case this is, essentially, Došen's [1989] semantics. The constructive minimal and the constructive case require some natural extensions, (cf. also [Thomason 1969], [López-Escobar 1972], [Gurevich 1977], [Rautenberg 1979], [van Dalen 1986] and [Akama 1988b]). Moreover, we shall argue to the effect that ***slomos*** in a certain sense provide an ***exhaustive*** picture of abstract information structures, thereby to a certain extent 'justifying' the rather general title of this investigation. We shall also briefly consider (linear) modalities and the paradigm of dynamic interpretation. A few applications of the monoid semantics are presented in an appendix.

9.1 Monoid models

We shall reproduce a version of Došen's [1989] groupoid semantics. This ***monoid semantics***, which will turn out to be adequate for the structural extensions of $MSPL$ and $ISPL$ considered earlier, is then generalized to the constructive minimal and the constructive case.

Definition 9.1 A semilattice-ordered monoid (*slomo*) is a structure $< I, \cdot, \cap, 1 >$ such that $1 \in I$, I is closed under the binary operations $\cdot$ and $\cap$, $\cdot$ is associative, $\cap$ is associative, commutative, and idempotent, for every $a \in I$, $a \cdot 1 = a = 1 \cdot a$, and $\cdot$

distributes over $\cap$ (i.e. for every $a, b, c \in I$: $a \cdot (b \cap c) = (a \cdot b) \cap (a \cdot c)$ and $(b \cap c) \cdot a = (b \cdot a) \cap (c \cdot a)$), cf. [Došen 1989, p. 43]. In a *slomo* the partial order $\leq$ is defined by $a \leq b$ iff $a = a \cap b$.

In Section 9.2 we shall develop an *informational interpretation* for the systems Ξ_Δ ($\Xi \in \{MSPL, ISPL, COSPL^-, COSPL\}$, $\Delta \subseteq \{\mathbf{P}, \mathbf{C}, \mathbf{C}', \mathbf{E}, \mathbf{E}', \mathbf{M}\}$), which is based on the following understanding of *slomos* $< I, \cdot, \cap, 1 >$:

- I is a set of information pieces;
- $\cdot$ is the *addition* of information pieces;
- $\cap$ is the *intersection* of information pieces;
- 1 is the *initial*, ideally the *empty* piece of information.

The defined relation $\leq$ can be interpreted as the 'possible development' (in the sense of 'possible expansion') or the 'possible prolongation' of information states or pieces. In this way the monoid semantics will link up with and generalize the semantical models which in Chapter 2 have been presented as information models.

Definition 9.2 (i) A structure $< I, \cdot, \cap, 1, v_0 >$ is called a monoid model for $MSPL$ iff $< I, \cdot, \cap, 1 >$ is a *slomo* and v_0 is a mapping from $PROP \cup \{\bot\}$ into 2^I such that for every $q \in PROP \cup \{\bot\}$ the following holds:

($\cap$ Heredity v_0) $\quad a_1 \cap a_2 \in v_0(q)$ iff ($a_1 \in v_0(q)$ and $a_2 \in v_0(q)$).

(ii) A monoid model $< I, \cdot, \cap, 1, v_0 >$ for $MSPL$ is a monoid model for $ISPL$ if for every $p \in PROP$ and every $a, b \in I$:

($\star$) $a \in v_0(\bot)$ implies $a \in v_0(p)$, $a \leq a \cdot b$, $a \leq b \cdot a$, $a \cdot a \leq a$, and $1 \leq a$.

Definition 9.3 The valuation v induced by a monoid model for $MSPL$ $< I, \cdot, \cap, 1, v_0 >$ is the function from the set of all L-formulas into 2^I inductively defined as follows (where $q \in PROP \cup \{\bot\}$):

$$
\begin{array}{lcl}
v(q) & = & v_0(q), \\
v(\mathbf{t}) & = & I, \\
v(\top) & = & \{a \mid 1 \leq a\}, \\
v(B/A) & = & \{a \mid (\forall b \in v(A))\, a \cdot b \in v(B)\}, \\
v(A \backslash B) & = & \{a \mid (\forall b \in v(A))\, b \cdot a \in v(B)\}, \\
v(A \circ B) & = & \{a \mid (\exists b_1 \in v(A))(\exists b_2 \in v(B))\, b_1 \cdot b_2 \leq a\}, \\
v(A \wedge B) & = & \{a \mid a \in v(A) \text{ and } a \in v(B)\}, \\
v(A \vee B) & = & \{a \mid (\exists b_1 \in v(A))(\exists b_2 \in v(B))\, b_1 \cap b_2 \leq a \text{ or } a \in v(A) \text{ or } a \in v(B)\}.
\end{array}
$$

Definition 9.4 (semantic consequence) Let $\mathcal{M} =< I, \cdot, \cap, 1, v_0 >$ be a monoid model for $MSPL$. If X is a non-empty sequence $A_1 \ldots A_n$, let $v(X) = v(A_1 \circ \ldots \circ A_n)$.

$$
X \to A \text{ holds (or is valid) in } \mathcal{M} \text{ iff } \begin{cases} v(X) \subseteq v(A) & \text{if } X \text{ is nonempty,} \\ 1 \in v(A) & \text{otherwise.} \end{cases}
$$

Lemma 9.5 (i) For every monoid model for $MSPL < I, \cdot, \sqcap, 1, v_0 >$, every a, $b \in I$, and every L-formula A:

($\sqcap$ Heredity v) $\quad a \sqcap b \in v(A)$ iff $(a \in v(A)$ and $b \in v(A))$.

(ii) For every monoid model for $ISPL < I, \cdot, \sqcap, 1, v_0 >$ and every L-formula A:

$v(\bot) \subseteq v(A)$.

PROOF By induction on the complexity of A, using $(\star)$ for (ii). Here is one case concerning (i): $A = (B/C)$:

$$\begin{array}{lll} & a \sqcap b \in v(B/C) & \\ \text{iff} & (\forall c \in v(C))(a \sqcap b) \cdot c \in v(B) & \\ \text{iff} & (\forall c \in v(C))(a \cdot c) \sqcap (b \cdot c) \in v(B) & \text{distrib. of} \cdot \text{over} \sqcap \\ \text{iff} & (\forall c \in v(C))(a \cdot c) \in v(B), & \\ & (\forall c \in v(C))(b \cdot c) \in v(B) & \text{ind. hyp.} \\ \text{iff} & a \in v(B/C) \text{ and } b \in v(B/C). \; \square & \end{array}$$

By the definition of $\leq$, it immediately follows from ($\sqcap$ Heredity v) that for every monoid model for $MSPL < I, \cdot, \sqcap, 1, v_0 >$, every a, $b \in I$, and every L-formula A:

(Heredity) $\quad$ if $a \leq b$, then $(a \in v(A)$ implies $b \in v(A))$.

Definition 9.6 (i) A structure $< I, \cdot, \sqcap, 1, v_0^+, v_0^- >$ is called a monoid model for $COSPL^-$ iff $< I, \cdot, \sqcap, 1 >$ is a *slomo* and v_0^+, v_0^- are mappings from $PROP \cup \{\bot, \sim \mathbf{t}, \sim \top\}$ into 2^I such that for every $q \in PROP \cup \{\bot, \sim \mathbf{t}, \sim \top\}$ the following holds:

($\sqcap$ Heredity v_0^+) $\quad a_1 \sqcap a_2 \in v_0^+(q)$ iff $(a_1 \in v_0^+(q)$ and $a_2 \in v_0^+(q))$.

($\sqcap$ Heredity v_0^-) $\quad a_1 \sqcap a_2 \in v_0^-(q)$ iff $(a_1 \in v_0^-(q)$ and $a_2 \in v_0^-(q))$.

(ii) A monoid model for $COSPL$ is a monoid model for $COSPL^- < I, \cdot, \sqcap, 1, v_0^+, v_0^- >$, where $v_0^+(\bot) = v_0^+(\sim \mathbf{t}) = v_0^+(\sim \top)$, $v_0^-(\bot) = I$, and for every $p \in PROP$ and every $a, b \in I$, $(\star)$ holds.

Definition 9.7 The valuation functions v^+, v^- induced by a monoid model for $COSPL^- < I, \cdot, \sqcap, 1, v_0^+, v_0^- >$ are the functions from the set of $L^\sim$-formulas into 2^I which are inductively defined as follows (where $q \in PROP \cup \{\bot, \sim \mathbf{t}, \sim \top\}$):

$$\begin{array}{lll} v^+(q) & = & v_0^+(q), \\ v^-(q) & = & v_0^-(q), \\ & & \\ v^+(\mathbf{t}) & = & I, \\ v^+(\top) & = & \{a \mid 1 \leq a\}, \\ & & \\ v^+(B/A) & = & \{a \mid (\forall b \in v^+(A)) a \cdot b \in v^+(B)\}, \\ v^-(B/A) & = & \{a \mid (\exists b_1 \in v^-(B))(\exists b_2 \in v^+(A)) b_1 \cdot b_2 \leq a\}, \end{array}$$

$$
\begin{aligned}
v^+(A \setminus B) &= \{a \mid (\forall b \in v^+(A)) b \cdot a \in v^+(B)\}, \\
v^-(A \setminus B) &= \{a \mid (\exists b_1 \in v^-(B))(\exists b_2 \in v^+(A)) b_2 \cdot b_1 \leq a\}, \\
\\
v^+(A \circ B) &= \{a \mid (\exists b_1 \in v^+(A))(\exists b_2 \in v^+(B)) b_1 \cdot b_2 \leq a\}, \\
v^-(A \circ B) &= \{a \mid (\exists b_1 \in v^-(A))(\exists b_2 \in v^-(B)) b_1 \cdot b_2 \leq a\}, \\
\\
v^+(A \wedge B) &= \{a \mid a \in v^+(A) \text{ and } a \in v^+(B)\}, \\
v^-(A \wedge B) &= \{a \mid (\exists b_1 \in v^-(A))(\exists b_2 \in v^-(B)) b_1 \cap b_2 \leq a \text{ or} \\
&\qquad a \in v^-(A) \text{ or } a \in v^-(B)\}, \\
\\
v^+(A \vee B) &= \{a \mid (\exists b_1 \in v^+(A))(\exists b_2 \in v^+(B)) b_1 \cap b_2 \leq a \text{ or} \\
&\qquad a \in v^+(A) \text{ or } a \in v^+(B)\}, \\
v^-(A \vee B) &= \{a \mid a \in v^-(A) \text{ and } a \in v^-(B)\}, \\
\\
v^+(\sim A) &= v^-(A), \\
v^-(\sim A) &= v^+(A).
\end{aligned}
$$

Thus, the definition of a valuation v in a monoid model for $MSPL$ resp. $ISPL$ agrees with the definition of v^+ in monoid models for $COSPL^-$ resp. $COSPL$. Moreover, the clauses $v^-(A)$ directly reflect the provable equivalences $(red1)$ and $(red2)$ in terms of $\rightleftharpoons^+$ listed in Chapter 6 (for instance $v^-(B/A)$ reflects $\sim (B/A) \rightleftharpoons^+ (\sim B \circ A)$). Therefore, to each $L^\sim$-formula A, one can find a (provably acceptance-equivalent) $L^\sim$-formula B such that $v^-(A) = v^+(\sim A) = v^+(B)$, and $\sim$ occurs in B only in front of propositional variables or constants. This fact can be used to simplify inductive proofs.

Definition 9.8 (semantic consequence) Let $\mathcal{M} = < I, \cdot, \cap, 1, v_0^+, v_0^- >$ be a monoid model for $COSPL^-$. If X is a non-empty sequence $A_1 \dots A_n$, let $v^+(X) = v^+(A_1 \circ \dots \circ A_n)$.

$$
X \rightarrow A \text{ holds (or is valid) in } \mathcal{M} \text{ iff } \begin{cases} v^+(X) \subseteq v^+(A) & \text{if } X \text{ is nonempty,} \\ 1 \in v^+(A) & \text{otherwise.} \end{cases}
$$

Lemma 9.9 (i) For every monoid model for $COSPL^- < I, \cdot, \cap, 1, v_0^+, v_0^- >$, every a, $b \in I$, and every $L^\sim$-formula A:

($\cap$ Heredity v^+) $a \cap b \in v^+(A)$ iff ($a \in v^+(A)$ and $b \in v^+(A)$);

($\cap$ Heredity v^-) $a \cap b \in v^-(A)$ iff ($a \in v^-(A)$ and $b \in v^-(A)$).

(ii) For every monoid model for $COSPL < I, \cdot, \cap, 1, v_0^+, v_0^- >$ and every $L^\sim$-formula A: $v^+(\perp) \subseteq v^+(A)$.

Proof By (simultaneous) induction on the complexity of A. For ($\cap$ Heredity v^-) it is enough to consider the cases where A is a propositional variable, $\perp$, **t**, or $\top$. □.

It can readily be verified that ($\cap$ Heredity v^+) resp. ($\cap$ Heredity v^-) implies that for every monoid model for $COSPL^- < I, \cdot, \cap, 1, v_0^+, v_0^- >$, every a, $b \in I$, and every $L^\sim$-formula A:

(Heredity $^+$) if $a \leq b$, then ($a \in v^+(A)$ implies $b \in v^+(A)$),

(Heredity $^-$) if $a \leq b$, then ($a \in v^-(A)$ implies $b \in v^-(A)$).

Theorem 9.10 (soundness) If $\vdash_\Xi X \to A$, then $X \to A$ holds in every monoid model for Ξ.

PROOF By induction on the complexity of proofs in Ξ. All cases are straightforward, except for ($\top \to$), ($\vee \to$) and ($\sim \wedge \to$). In the latter two cases one has to make use of both ($\cap$ Heredity v^+) and (Heredity $^+$) (cf. [Došen 1989, p. 48]). Let us here consider ($\top \to$) and ($\sim \wedge \to$). Let $\mathcal{M} = < I, \cdot, \cap, 1, v_0 >$ resp. $\mathcal{M} = < I, \cdot, \cap, 1, v_0^+, v_0^- >$ be any monoid model for Ξ. ($\top \to$): It is enough to show that $v^+(\top \circ A) \subseteq v^+(A)$ and $v^+(A \circ \top) \subseteq v^+(A)$. Consider the latter. Note that if $a \leq b$, then $c \cdot a \leq c \cdot b$, for every a, b, $c \in I$. Suppose $c \in v^+(A \circ \top)$. Then $(\exists b_1 \in v^+(A))\ (\exists b_2 \in v^+(\top))\ b_1 \cdot b_2 \leq c$. Now, $1 \leq b_2$. Therefore $b_1 = b_1 \cdot 1 \leq b_1 \cdot b_2$. By transitivity of $\leq$, $b_1 \leq c$. Hence, by (Heredity $^+$), $c \in v^+(A)$. ($\sim \wedge \to$): Let $C = C_1 \circ \ldots \circ C_n$, $D = D_1 \circ \ldots \circ D_n$, and suppose that $v^+(C \sim AD) \subseteq v^+(E)$, $v^+(C \sim BD) \subseteq v^+(E)$. Then $v^+(C \sim A) \subseteq v^+((E/D))$ and $v^+(C \sim B) \subseteq v^+((E/D))$. Hence $(\forall a \in v^+(C))\ (\forall b \in v^+(\sim A))\ a \cdot b \in v^+((E/D))$, $(\forall a \in v^+(C))\ (\forall b \in v^+(\sim B))\ a \cdot b \in v^+((E/D))$. Therefore,

$$
\begin{array}{lll}
& a \in v^+(C),\ b_1 \in v^+(\sim A),\ b_2 \in v^+(\sim B),\ b_1 \cap b_2 \leq b & \\
\text{only if} & a \cdot b_1 \in v^+((E/D)) \text{ and } a \cdot b_2 \in v^+((E/D)) & \\
\text{only if} & (a \cdot b_1) \cap (a \cdot b_2) \in v^+((E/D)) & (\cap \text{ Heredity } v^+) \\
\text{only if} & a \cdot (b_1 \cap b_2) \in v^+((E/D)) & \\
\text{only if} & (a \cdot b) \in v^+((E/D)) & (\text{Heredity }^+).
\end{array}
$$

Since also ($a \in v^+(C)$ and ($b \in v^+(\sim A)$ or $b \in v^+(\sim B)$)) only if $a \cdot b \in v^+((E/D))$, we obtain

$$
\begin{array}{ll}
& a \in v^+(C) \\
\text{only if} & (\forall b \in v^+(\sim A \vee \sim B))\ a \cdot b \in v^+((E/D)) \\
\text{iff} & (\forall b \in v^+(\sim (A \wedge B))\ a \cdot b \in v^+((E/D)) \\
\text{only if} & v^+(C \sim (A \wedge B) D) \subseteq v^+(C). \ \Box
\end{array}
$$

Note that also in monoid models for *ISPL* and *COSPL* inconsistent pieces of information are not forbidden; $v_0^+(p) \cap v_0^-(p) \neq \emptyset$ is not excluded. Therefore, although the *ex falso* rule ($\bot \to$) is valid, *ex contradictione* is not.

Using ($\cap$ Heredity v), Došen [1989, p. 52 f.] proves a number of correspondences between structural rules of inference and conditions on *slomos* in the sense that a given structural rule R is validity preserving in a monoid model $\mathcal{M}$ iff the condition on *slomos* corresponding to R is satisfied by the *slomo* on which $\mathcal{M}$ is based. In the present context we have the following correspondences:

	for every $a, b \in I$	
P	$a \cdot b \leq b \cdot a$	
C	$a \cdot a \leq a$	
C′	$a \cdot b \cdot a \leq a \cdot b,$	$a \cdot b \cdot a \leq b \cdot a$
E	$a \leq a \cdot a$	
E′	$a \cdot b \leq a \cdot b \cdot a,$	$b \cdot a \leq a \cdot b \cdot a$
M	$1 \leq a$	

Table 8.1: Structural rules and conditions on *slomos*.

Note that the behaviour of $\sim$ is not reflected in these structural conditions on *slomos*; it is completely captured by the valuations v^- and v^+. By the above correspondences, Ξ_Δ is sound wrt the class of monoid models for Ξ whose underlying *slomos* satisfy the conditions that correspond to the rules in Δ. Let us call this class of models M_{Ξ_Δ}.[1]

9.2 Informational interpretation

By itself, the monoid semantics of the previous section might convey the impression of being just a technical tool without very much explanatory value and intuitive appeal. In this section we shall, however, develop an informational interpretation (in the sense of Appendix 2.5) for the systems Ξ_Δ that is based on the above mentioned understanding of *slomos* $< I, \cdot, \sqcap, 1 >$, i.e., I is a set of information pieces or information states, $\cdot$ is the addition of information pieces, $\sqcap$ is the intersection of information pieces, 1 is the initial, ideally the empty piece of information, and $\leq$ is the prolongation resp. development of information pieces resp. states. Of course, this reading is illuminating only if we are willing to attach some explanatory power to the notions of addition and intersection of information pieces. We claim that under the suggested reading the properties which in *slomos* are postulated for $\cdot$, $\sqcap$, and 1 are intuitively plausible. Or, to state it the other way around, if we have a set I of information pieces including one initial piece of information together with one addition and one intersection operation on I, then it is plausible to assume that these components should form a *slomo*.

What can we say about the evaluation of formulas in monoid models? The valuation function v resp. v^+ in monoid models for $MSPL$ resp. $COSPL^-$ specifies truth conditions; the valuation v^- in monoid models for $COSPL^-$ in addition specifies falsity conditions, where falsity is falsity in the sense of refutation. Thus, in contrast to the minimal intuitionistic and the intuitionistic case, in the minimal constructive and the constructive systems truth and falsity are regarded as prima facie independent notions.

[1] There are some minor differences between our presentation and Došen's semantics in [1989] which it might be useful to briefly point out. Although **P** is absent, Došen considers only one implication sign, viz. /, which he denotes by $\rightarrow$ instead of $\leftarrow$. Moreover, Došen has only one verum constant in his language, viz. $\top$. Whereas $v(\top) = \{a \mid 1 \leq a\}$, $v(\mathbf{t}) = W$. This ensures that $\top$ entails **t** but not conversely and, moreover, that both are interderivable, if the condition on *slomos* corresponding to **M** is satisfied. In [1988, p. 366] Došen explains that $\top$ in his basic axiomatic system behaves as an arbitrary propositional variable. This is, however, not reflected in the groupoid semantics; since $\top$ is not evaluated by v_0. The treatment of $\top$ as a propositional variable explains why 1 is not assumed to be a neutral element wrt

Finally, Došen considers extensions of his base logic by $\vdash \bot \rightarrow A$ instead of $(\bot \rightarrow)$, which has as a consequence that applications of (*cut*) cannot be eliminated, if **M** is absent.

Whereas for the minimal logics inconsistent pieces of information are admitted, this is not the case for structural extensions of *ISPL* and *COSPL*. Now, do the valuation clauses emerge as plausible under the given interpretation of *slomos*? Let us first consider the valuations v and v^+. The evaluation of elements from *PROP* resp., in the minimal cases, $PROP \cup \{\bot\}$ or $PROP \cup \{\bot, \sim \mathbf{t}, \sim \top\}$, is unproblematic. ($\cap$ Heredity v_0), ($\cap$ Heredity v_0^+), and ($\cap$ Heredity v_0^-) are natural requirements. If a formula A is true on the strength of the intersection $a \cap b$ of information pieces a, b, then A should be true on the strength of both pieces, and conversely. If $\bot$, $\sim \mathbf{t}$ and $\sim \top$ are treated as falsum constants, then any formula should be derivable from premises containing at least one of these falsum constants. This is guaranteed by ($\star$) and the fact that ($\uparrow$ /) and ($\uparrow$ \) preserve validity. The evaluation of $\mathbf{t}$ and $\top$ is without doubt reasonable: we may distinguish between a verum constant which is true at every information state and another verum constant which is true at every information state into which the initial piece of information may develop. The clauses for / and \ are just directional versions of Urquhart's [1972] truth definition. Moreover, it is rather natural to say that if A is true at information state b_1 and B is true at information state b_2, then $(A \circ B)$, which is a conjunction in the sense of juxtaposition, is true at every information piece into which $b_1 \cdot b_2$ may develop. The case of $\wedge$ is again unproblematic. In the case of $\vee$ it makes perfectly good sense to require that $(A \vee B)$ is true not only at pieces of information a at which A is true or at which B is true but also at pieces of information which prolong the intersection of pieces of information b_1 and b_2 such that A is true at b_1 and B is true at b_2. Thus, $(A \vee B)$ should also be true at information pieces which prolong so to speak the common content of information pieces b_1, b_2 with A true at b_1 and B true at b_2.[2] Finally, the evaluation of $\sim A$ by means of v^+ is intuitively convincing. We have that $\sim A$ is true at a piece of information a iff A is false at a. Turning to v^- we can thus say that the definition of $v^-(\sim A)$ is intuitively sound. In general the definition of v^- can be justified by the naturalness of the provable equivalences (*red*1) and (*red*2) (see the discussion in Chapter 6). Moreover the definition of semantic consequence is in accordance with what we have said so far. The information states into which the initial (or empty) piece of information 1 may develop should take precedence over the set of all information pieces.

In order to show that the present interpretation of monoid models is in fact informational according to the criteria suggested earlier, we have to provide a model which can arguably be talked about as the intended model under the given interpretation and which is a canonical model for the logic in question. Now, the following assumptions seem to be natural: (i) Think of information pieces as finite sequences of formula occurrences, since in our basic calculi the databases are juxtapositions of such finite sequences. (ii) Identify those pieces of information which are interderivable (identifying $A_1 \ldots A_n$ with $A_1 \circ \ldots \circ A_n$, if $n > 1$). This is enough from the point of view of deductive information processing, although the representatives need not be synonymous in the sense of being intersubstitutable in all deductive contexts: if A and B are interderivable, by (*cut*) we have in Ξ_Δ, $\vdash XAY \rightarrow C$ iff $\vdash XBY \rightarrow C$ and $\vdash X \rightarrow A$ iff $X \rightarrow B$. The formulas interderivable with $\top$ can then be viewed as representing the empty piece of informa-

[2] Došen [1989, p. 45] motivates the evaluation clause for disjunction by pointing out an analogous clause in Birkhoff's and Frink's representation of lattices by sets.

tion, since we have that $\vdash_{\Xi_\Delta} \rightarrow A$ iff $\vdash_{\Xi_\Delta} \top \rightarrow A$. Next, let a and b be two pieces of information, and let A resp. B be a representative of a resp. b. The addition $a \cdot b$ of a and b should be the equivalence class of $(A \circ B)$ wrt interderivability, and the intersection $a \cap b$ of a and b should be the equivalence class of $(A \vee B)$ wrt interderivability. These considerations naturally lead us to the following definition of intended models.

Definition 9.11 Let $\overset{\circ}{X} = A_1 \circ \ldots \circ A_n$, if $X = A_1 \ldots A_n\, (n > 1)$, and let $\overset{\circ}{X} = A$, if $X = A$. For every formula A, the equivalence class of A modulo $\leftrightarrow$ will be denoted by $\mid A \mid$. The intended model $\mathcal{M}_{\Xi_\Delta} =< I, \cdot, \cap, 1, v_0 >$ resp. $\mathcal{M}_{\Xi_\Delta} =< I, \cdot, \cap, 1, v_0^+, v_0^- >$ for Ξ_Δ is defined as follows, where $q \in PROP \cup \{\bot\}$ resp. $PROP \cup \{\bot, \sim \mathbf{t}, \sim \top\}$:

- $I = \{|\overset{\circ}{X}| \mid X$ is a non-empty sequence of formulas occurrences $\}$;
- $|\overset{\circ}{X}_1| \cdot |\overset{\circ}{X}_2| = |\overset{\circ}{X}_1 \circ \overset{\circ}{X}_2|$;
- $|\overset{\circ}{X}_1| \cap |\overset{\circ}{X}_2| = |\overset{\circ}{X}_1 \vee \overset{\circ}{X}_2|$;
- $1 =\mid \top \mid$;
- $v_0(q) = \{|\overset{\circ}{X}| \mid \vdash_{\Xi_\Delta} X \rightarrow q\}$;
- $v_0^+(q) = v_0(q)$, with $v_0^+(\bot) = v_0^+(\sim \mathbf{t}) = v^+(\sim \top)$, if $\Xi = COSPL$;
- $v_0^-(q) = \{|\overset{\circ}{X}| \mid \vdash_{\Xi_\Delta} X \rightarrow \sim q\}$, with $v_0^-(\bot) = I$, if $\Xi = COSPL$.

This construction clearly has an algebraic twist; note, however, that it does ***not*** yield the so-called Lindenbaum-algebra for Ξ_Δ. In the constructive case e.g. there are no algebraic operations corresponding to $\rightleftharpoons$, $/$, $\backslash$, $\mathbf{t}$, and $\sim$ (cf. Rasiowa's [1974, p. 68] quasi-pseudo-Boolean algebras for **N**): the present semantics is close to syntax, but it is not 'syntax in disguise'.

Lemma 9.12 $\mathcal{M}_{\Xi_\Delta}$ is in fact a monoid model for Ξ.

PROOF Obviously, $1 \in I$ and I is closed under $\cdot$ and $\cap$. Associativity of $\cdot$ and associativity, commutativity, and idempotence of $\cap$ are immediate. To see that 1 is a neutral element wrt $\cdot$, observe that $\vdash_{\Xi_\Delta} (\top \circ A) \leftrightarrow A$ and $\vdash_{\Xi_\Delta} (A \circ \top) \leftrightarrow A$. Distributivity of $\cdot$ over $\cap$ follows by (†) (Chapter 3). Eventually, we have to check ($\cap$ Heredity v_0^+) and ($\cap$ Heredity v_0^-). We check the latter property, to check the former is completely analogous:

$$
\begin{array}{lll}
 & |\overset{\circ}{X}_1| \in v_0^-(q) \text{ and } |\overset{\circ}{X}_2| \in v_0^-(q) & \\
\text{iff} & \vdash X_1 \rightarrow\sim q \text{ and } \vdash X_2 \rightarrow\sim q & \text{Def. } v_0^- \\
\text{iff} & \vdash \overset{\circ}{X}_1 \vee \overset{\circ}{X}_2 \rightarrow\sim q & (\circ \rightarrow), (\circ \uparrow), (\vee \rightarrow), (\vee \uparrow) \\
\text{iff} & |\overset{\circ}{X}_1 \vee \overset{\circ}{X}_2| \in v_0^-(q) & \text{Def. } v_0^- \\
\text{iff} & |\overset{\circ}{X}_1| \cap |\overset{\circ}{X}_2| \in v_0^-(q) & \text{Def. } \cap .
\end{array}
$$

It is straightforward to verify that in $\mathcal{M}_{ISPL_\Delta}$ and $\mathcal{M}_{COSPL_\Delta}$ for every $|\mathring{X}|$, $|\mathring{Y}| \in I$ and every $p \in PROP$, $|\mathring{X}| \in v_0(\bot)$ implies $|\mathring{X}| \in v_0(p)$, $|\mathring{X}| \leq |\mathring{X}| \cdot |\mathring{Y}|$, $|\mathring{X}| \leq |\mathring{Y}| \cdot |\mathring{X}|$, $|\mathring{X}| \cdot |\mathring{X}| \leq |\mathring{X}|$, and $1 \leq |\mathring{X}|$ □.

We shall now show that the intended monoid model for Ξ_Δ is a canonical model for Ξ_Δ and therefore our interpretation of monoid models is informational according to the constraints suggested in Chapter 2.

Lemma 9.13 (Truthlemma) For every $|\mathring{X}| \in I$ and every L-formula resp. $L^\sim$-formula A, the intended model $\mathcal{M}_{\Xi_\Delta}$ satisfies:

$|\mathring{X}| \in v(A)$ resp. $v^+(A)$ iff $\vdash_{\Xi_\Delta} X \to A$.

PROOF By induction on the complexity of A.

- $A = q \in PROP \cup \{\bot, \sim \mathbf{t}, \sim \top\}$: by the definition of v_0 resp. v_0^+.

- $A = \mathbf{t}$: $|\mathring{X}| \in v(\mathbf{t})$ iff $|\mathring{X}| \in I$ iff $\vdash X \to \mathbf{t}$.

- $A = \top$:

$$
\begin{array}{lll}
 & |\mathring{X}| \in v(\top) & \\
\text{iff} & 1 \leq |\mathring{X}| & \\
\text{iff} & |\top| \cap |\mathring{X}| = |\top| & \text{def. } \leq \\
\text{iff} & |\top \vee \mathring{X}| = |\top| & \text{def. } \cap \\
\text{iff} & \vdash X \to \top & (\vee \uparrow), (id), (\vee \to).
\end{array}
$$

- $A = \sim B$: It is enough to consider the case where B is a propositional variable or constant. Then the claim holds by the definition of v_0^-.

- $A = (B \setminus C)$:

$$
\begin{array}{lll}
 & |\mathring{X}| \in v(B \setminus C) & \\
\text{iff} & (\forall |\mathring{Y}| \in v(B))\ |\mathring{Y}| \circ |\mathring{X}| \in v(C) & \\
\text{iff} & (\forall |\mathring{Y}| \in v(B))\ |\mathring{Y} \circ \mathring{X}| \in v(C) & \\
\text{iff} & \forall Y (\text{if } \vdash Y \to B, \text{ then } \vdash YX \to C) & \text{ind. hyp.} \\
\text{iff} & \vdash BX \to C & (cut), (id) \\
\text{iff} & \vdash X \to (B \setminus C) & (\uparrow \setminus), (\to \setminus).
\end{array}
$$

- $A = (C/B)$: analogous to the previous case.

- $A = (B \circ C)$:

$$
\begin{array}{lll}
& \vdash X \to (B \circ C) & \\
\text{iff} & \forall A (\text{if } \vdash B \circ C \to A, \text{ then } \vdash X \to A) & (cut), (id) \\
\text{iff} & \forall A (\text{if } \vdash BC \to A, \text{ then } \vdash X \to A) & (\circ \uparrow), (\circ \to) \\
\text{iff} & \exists Y_1 \exists Y_2 (\vdash Y_1 \to B \text{ and } \vdash Y_2 \to C \text{ and} & \\
& \forall A (\text{if } \vdash Y_1 Y_2 \to A, \text{ then } \vdash X \to A)) & (id) \\
\text{iff} & (\exists |\mathring{Y}_1| \in v(B))(\exists |\mathring{Y}_2| \in v(C)) \vdash \mathring{X} \to \mathring{Y}_1 \circ \mathring{Y}_2 & \text{ind. hyp., } (cut) \\
\text{iff} & (\exists |\mathring{Y}_1| \in v(B))(\exists |\mathring{Y}_2| \in v(C)) \; |\mathring{Y}_1 \circ \mathring{Y}_2| \leq |\mathring{X}| & \text{def. } \leq, (\vee \uparrow), (\to \vee) \\
\text{iff} & |\mathring{X}| \in v(B \circ C). &
\end{array}
$$

- $A = (B \wedge C)$: use the induction hypothesis.
- $A = (B \vee C)$:

$$
\begin{array}{lll}
& |\mathring{X}| \in v(B \vee C) & \\
\text{iff} & (\exists |\mathring{Y}_1| \in v(B))(\exists |\mathring{Y}_2| \in v(C)) (|\mathring{Y}_1 \vee \mathring{Y}_2| \cap |\mathring{X}| = & \\
& |\mathring{Y}_1 \vee \mathring{Y}_2| \text{ or } |\mathring{X}| \in v(B) \text{ or } |\mathring{X}| \in v(C)) & \\
\text{iff} & \exists \mathring{Y}_1 \exists \mathring{Y}_2 (\vdash \mathring{Y}_1 \to B \text{ and } \mathring{Y}_2 \to C \text{ and} & \\
& \vdash (\mathring{Y}_1 \vee \mathring{Y}_2) \vee \mathring{X} \leftrightarrow \mathring{Y}_1 \vee \mathring{Y}_2) \text{ or} & \\
& \vdash \mathring{X} \to B \text{ or } \mathring{X} \to C & \text{ind. hyp., def. } \cap \\
\text{iff} & \vdash \mathring{X} \to (B \vee C) & (\vee \uparrow), (\vee \to), (cut), (id) (\to \vee) \\
\text{iff} & \vdash X \to (B \vee C) & (\circ \uparrow) (\circ \to). \;\square
\end{array}
$$

With the Truthlemma in our hands, we are in a position to prove completeness.

Theorem 9.14 (completeness) If $X \to A$ holds in every monoid model from $\mathsf{M}_{\equiv_\Delta}$, then $\vdash_{\equiv_\Delta} X \to A$.

PROOF Suppose that $\not\vdash_{\equiv_\Delta} X \to A$. By the Truthlemma, this is the case iff in $\mathcal{M}_{\equiv_\Delta}$ we have that $|\mathring{X}| \notin v^+(A)$. But this implies that $v^+(\mathring{X}) \not\subseteq v(A)$, since $|\mathring{X}| \in v(\mathring{X})$. It remains to be shown that in each case the underlying *slomo* of the intended model satisfies the conditions which correspond to the rules in Δ, i.e. $\mathcal{M}_{\equiv_\Delta} \in \mathsf{M}_{\equiv_\Delta}$. But this is a completely straightforward matter. Consider by way of example the case of the structural rule **E**. It has to be shown that $| A | \leq | A | \cdot | A |$, i.e. $| A | \cap | A \cdot A | = | A |$. Now, using **E** it can easily be seen that $\vdash_{\equiv_\Delta} A \vee (A \circ A) \leftrightarrow A$. □

Next, consider the following conception of an intended monoid model. Think of a piece of information as the deductive closure of a finite sequence of formula occurrences. The intersection of information pieces should then be nothing but set intersection, and the addition $a_1 \cdot a_2$ of two pieces of information a_1, a_2 with finite representations X_1,

X_2 should be the deductive closure of X_1X_2, i.e. the deductive closure of the juxtaposition of their representations. The empty piece of information would be represented by the set of all theorems, i.e. the deductive closure of the empty sequence. Where these considerations lead us is Došen's [1989] construction of a canonical monoid model:

Definition 9.15 The canonical monoid model $\mathcal{M}'_{\Xi_\Delta} = < I', \cdot', \cap', 1', v_0 >$ resp. $\mathcal{M}'_{\Xi_\Delta} = < I', \cdot', \cap', 1', v_0^+, v_0^- >$ for Ξ_Δ is defined as follows, where $q \in PROP \cup \{\bot\}$ resp. $PROP \cup \{\bot, \sim \mathbf{t}, \sim \top\}$:

- $I' = \{a \mid \exists X, a = \{A \mid \vdash_{\Xi_\Delta} X \to A\}\}$;
- if $a_1 = \{A \mid \vdash_{\Xi_\Delta} X_1 \to A\}$ and $a_2 = \{A \mid \vdash_{\Xi_\Delta} X_2 \to A\}$, then $a_1 \cdot' a_2 = \{A \mid \vdash_{\Xi_\Delta} X_1X_2 \to A\}$;
- $\cap'$ is set intersection;
- $1' = \{A \mid \vdash_{\Xi_\Delta} \to A\}$
- $v_0(q) = \{a \in I \mid q \in a\}$;
- $v_0^+(q) = v_0(q)$, with $v_0^+(\bot) = v_0^+(\sim \mathbf{t}) = v^+(\sim \top)$, if $\Xi = COSPL$;
- $v_0^-(q) = \{a \in I \mid \sim q \in a\}$, with $v_0^-(\bot) = I$, if $\Xi = COSPL$.

It can easily be shown that $\mathcal{M}'_{\Xi_\Delta}$ is in fact a monoid model for Ξ and that $\mathcal{M}'_{\Xi_\Delta} \in \mathsf{M}_{\Xi_\Delta}$. Assume for example $\mathbf{C} \in \Delta$. Suppose that $a = \{A \mid\vdash X \to A\}$, $X = A_1 \ldots A_n$. Since

$$\dfrac{X \to A_1 \circ \ldots \circ A_n \qquad \dfrac{\dfrac{A_1 \ldots A_n A_1 \ldots A_n \to A}{A_1 \circ \ldots \circ A_n A_1 \circ \ldots \circ A_n \to A}}{A_1 \circ \ldots \circ A_n \to A}}{X \to A}$$

$\{A \mid\vdash XX \to A\} \subseteq \{A \mid\vdash X \to A\}$, i.e. $a \cdot a \leq a$.

As it turns out, both constructions of canonical models are isomorphic and can therefore be identified:

Observation 9.16 There exists an isomorphism between the underlying *slomos* of $\mathcal{M}'_{\Xi_\Delta}$ and $\mathcal{M}_{\Xi_\Delta}$.

PROOF The function $h : I' \longrightarrow I$ defined by $h(\{A \mid\vdash_{\Xi_\Delta} X \to A\}) = |\mathring{X}|$ is such an isomorphism. $1-1$: Suppose that $a = \{A \mid\vdash_{\Xi_\Delta} X \to A\} \neq b = \{B \mid\vdash_{\Xi_\Delta} Y \to B\}$, but $h(a) = h(b)$. Then $\vdash_{\Xi_\Delta} \mathring{X} \leftrightarrow \mathring{Y}$ and by (cut) and $(\uparrow \circ)$, $a = b$. Onto: obvious. The homomorphism property is easy to establish. Let $a = \{A \mid\vdash_{\Xi_\Delta} X \to A\}$, $b = \{B \mid\vdash_{\Xi_\Delta} Y \to B\}$. $h(a \cdot' b) = |\mathring{X} \circ \mathring{Y}| = h(a) \cdot h(b)$. $h(a \cap' b) = \{A \mid\vdash_{\Xi_\Delta} X \to A\} \cap \{B \mid\vdash_{\Xi_\Delta} Y \to B\} = \{C \mid\vdash_{\Xi_\Delta} X \to C$ and $\vdash_{\Xi_\Delta} Y \to C\} = \{C \mid\vdash_{\Xi_\Delta} \mathring{X} \vee \mathring{Y} \to C\}$, by $(\uparrow \circ)$, $(\vee \to)$. Thus $h(a \cap' b) = |\mathring{X} \vee \mathring{Y}| = |\mathring{X}| \cap |\mathring{Y}| = h(A) \cap h(B)$. □

In conclusion we may say that the monoid semantics provides an informational interpretation for a broad range of substructural propositional logics, including the limiting cases $\mathbf{N}^-, \mathbf{N}, MPL$, and IPL. Moreover, different conceptions of deductive information processing within one family of formal systems naturally correspond to different conceptions of *slomos* as abstract information structures.

9.3 Extension with (linear) modalities

Although, as explained in a footnote to Chapter 2, the present investigation is concerned with the 'subjective understanding of logic' as opposed to the 'objective' or 'external' understanding which involves modal belief operators, it might be interesting to give a short outlook on possible treatments of modal operators. We shall briefly consider the so-called 'exponential' ! ("of course") of Girard's intuitionistic linear logic (see [Girard 1987], [Avron 1988], Troelstra [1992]).

The rules for Girard's ! can be divided into two groups, viz. rules for introducing ! into premises and conclusions, and rules mimicking the structural rules **C** and **M** for formulas prefixed by an occurrence of !. We may also add rules mimicking **P**, **C′**, **E**, and **E′**:

$$\begin{array}{ll} (\rightarrow !) & XAY \rightarrow B \vdash X!AY \rightarrow B; \\ (! \rightarrow) & !A_1 \ldots !A_n \rightarrow A \vdash !A_1 \ldots !A_n \rightarrow !A; \\ (!\mathbf{P}) & X!A!BY \rightarrow C \vdash X!B!AY \rightarrow C; \\ (!\mathbf{C}) & X!A!AY \rightarrow B \vdash X!AY \rightarrow B; \\ (!\mathbf{C}') & X!AY!AZ \rightarrow B \vdash XY!AZ \rightarrow B, \\ & X!AY!AZ \rightarrow B \vdash X!AYZ \rightarrow B; \\ (!\mathbf{E}) & X!AY \rightarrow B \vdash X!A!AY \rightarrow B; \\ (!\mathbf{E}') & XY!AZ \rightarrow B \vdash X!AY!AZ \rightarrow B, \\ & X!AYZ \rightarrow B \vdash X!AY!AZ \rightarrow B; \\ (!\mathbf{M}) & XY \rightarrow B \vdash X!AY \rightarrow B. \end{array}$$

It is well-known that with the rules $(\rightarrow !)$ and $(! \rightarrow)$, ! is an **S4**-modality. That is to say ! can be interpreted by means of a reflexive and transitive binary relation $R^!$, using the standard Kripkean truth definition

(!) $v(!A) = \{a \in I \mid (\forall b \in I)(aR^!b \text{ implies } b \in v(A))\}$.

$R^!$ may be understood as the 'accessibility relation' between information states.

Definition 9.17 A *!-slomo* is a structure $< I, R^!, \cdot, \cap, 1 >$, where $< I, \cdot, \cap, 1 >$ is a *slomo*, $R^!$ is a reflexive and transitive binary relation on I, and $\leq \subseteq R^!$.

In other words, every expansion is an accessible information state, but the converse is not true in general.

Definition 9.18 Let $!MSPL_\Delta$ resp. $!ISPL_\Delta$ denote the result of adding the above rules for ! to $MSPL_\Delta$ resp. $ISPL_\Delta$.

Definition 9.19 !-monoid models for $!MSPL$ resp. $!ISPL$ are defined exactly as monoid models for $MSPL$ resp. $ISPL$ except that they are based on !-*slomos* and ! is interpreted according to (!).

Lemma 9.20 For every !-monoid model $< I, R^!, \cdot, \cap, 1, v_0 >$ for $!MSPL$ resp. $!ISPL$, every a, $b \in I$, and every L-formula A:

(Heredity) if $a \leq b$, then ($a \in v(A)$ implies $b \in v(A)$).

PROOF By induction on A; we consider $A =!B$. Suppose $a \leq b$. If $a \in v(!B)$, then $(\forall b \in I)$ ($aR^!b$ implies $b \in v(B)$). By transitivity of $R^!$, $(\forall b \in I)$ ($aR^!b$ implies $b \in v(!B)$). Since $\leq \subseteq R^!$, $(\forall b \in I)$ ($a \leq b$ implies $b \in v(!B)$). □

Observation 9.21 Each structural inference rule R listed in the following table is validity preserving in a !-monoid model $\mathcal{M}$ for $!MSPL$ resp. $!ISPL$, if the condition on !-*slomos* associated with R is satisfied by the !-*slomo* on which $\mathcal{M}$ is based:

	for every a, b, c, $d \in I$, for every A
!P	$a \in v(!A)$ and $b \in v(!B)$ implies $a \cdot b \leq b \cdot a$
!C	$a \in v(!A)$ implies $a \cdot a \leq a$
!C′	$a \in v(!A)$ implies $a \cdot b \cdot a \leq a \cdot b$, $a \in v(!A)$ implies $a \cdot b \cdot a \leq b \cdot a$
!E	$a, b \in v(!A)$ and $a \cdot b \leq c$ implies $a \leq c$, $b \leq c$
!E′	$a, b \in v(!A)$ and $a \cdot c \cdot b \leq d$ implies $a \cdot c \leq d$, $a, b \in v(!A)$ and $a \cdot c \cdot b \leq d$ implies $c \cdot b \leq dc$
!M	$a \in v(!A)$ implies $1 \leq a$

Table 8.2: Structural rules and conditions on !-*slomos*.

These conditions are 'mixed'; they combine structural properties and valuations. Therefore this semantics is probably not restrictive enough to count as very illuminating. We leave it as an open question whether the soundness results which follow from the above observation can be supplemented by completeness theorems for the respective model classes.[3] Providing a clear *structural* characterization for Girard's modalities is one of the main open problems in the area.

Note that the rule $(\rightarrow!)$ fails to be an instantiation of the earlier schemata (I) for introducing connectives into premises (see Chapters 4 and 7). This again points to a certain narrowness of (I). Note also that the methodology which determines rules for introductions on the lhs of $\rightarrow$ *does* apply to $(! \rightarrow)$. Consider an application of (cut) such that the last steps in proving the premise sequents of this application introduce ! as the main connective of the cut-formula:

$$\dfrac{\dfrac{\dfrac{\Pi_1}{!A_1 \ldots !A_n \rightarrow A}}{!A_1 \ldots !A_n \rightarrow !A} \quad \dfrac{\Pi_2}{X!AZ \rightarrow B}}{X!A_1 \ldots !A_n Z \rightarrow B.}$$

[3]A completeness proof for the propositional modal logic **S4** based on (the propositional part of) Nelson's systems $\mathbf{N}^-$ and **N** but in a language without ◇ is sketched in [Routley 1974].

In order to remove this application of (*cut*) by an application of (*cut*) with smaller degree, now also counting occurrences of !, one in fact ends up with (! →):

$$\frac{\dfrac{\Pi_1}{!A_1 \ldots !A_n \to A} \quad \dfrac{\Pi_2}{XAZ \to B}}{X!A_1 \ldots !A_n Z \to B.}$$

Thus, one obvious direction into which the schemata (I) could be liberalized would be to allow more than one occurrence of the connective that is introduced.

Another obvious question in this connection concerns the refutability conditions of $!A$. A natural idea is to consider ! as the dual of a possibility operator $\Diamond$ with evaluation clauses

($\Diamond^+$) $v^+(\Diamond A) = \{a \in I \mid (\exists b \in I)\ (aR^!b \text{ and } b \in v^+(A))\}$,
($\Diamond^-$) $v^-(\Diamond A) = \{a \in I \mid (\forall b \in I)\ (aR^!b \text{ implies } b \in v^+(\sim A))\}$,

such that one obtains

($!^+$) $v^+(!A) = v(!A)$,
($!^-$) $v^-(!A) = \{a \in I \mid (\exists b \in I)\ (aR^!b \text{ and } b \in v^+(\sim A))\}$.

Note that distinguishing between truth and falsity conditions in modal logic is something of interest by itself, in particular if we think of an epistemic reading of modal operators.

9.4 Dynamics of interpretation

Talking about inference as the dynamics of information processing, we should also briefly relate our informational interpretation by means of monoid models to the recent paradigm of *dynamic interpretation*. We shall consider dynamic predicate logic DPL [Groenendijk & Stokhof 1991], since DPL's underlying philosophy of meaning has been clearly worked out and applied. DPL has been developed by Groenendijk and Stokhof as a compositional semantics for pronominal anaphora and is primarily designed for a translation of natural language texts into the language of *first-order* logic. Therefore the comparison to the monoid semantics for substructural *propositional* logics must remain partial and will cover only one fundamental idea behind DPL, viz. the *philosophy of meaning* on which DPL is based and which can be put into a slogan as follows:

> The meaning of a declarative sentence is the contribution
> of this sentence to the transformation of information states.

Thus, suppose someone is in information state b, and happens to understand a declarative sentence A (which is uttered or which she can read). Understanding A brings her into another information state b', and if the information conveyed by A is new, b' will differ from b. In other words, the meaning of a declarative sentence may change information states or, in more static words, relate such states. Of course, the possible output of the alteration process depends on the input, but not necessarily as a function.[4] The interpretation of a sentence A in DPL therefore becomes a binary *relation* $\|A\|$ between

[4]Cf., however, e.g. [Belnap 1977] and [Pearce & Rautenberg 1991].

possible input and output states (which in *DPL* are represented as assignments of objects to individual variables). Let us consider a few important examples. *DPL* assumes as interpretation of the full stop '.' (between declarative sentences) the obvious relational interpretation of $\circ$, viz. the composition of binary relations (see [van Benthem 1989]). This is quite natural in view of our interpretation of $\circ$ as a text-forming operation of Categorial Grammar in Chapter 4. The meaning $\| (A \circ B) \|$ of $(A \circ B)$[5] in *DPL* thus is $\{< a,b >| \exists c(< a,c >\in \|A\| \text{ and } < c,b >\in \|B\|)\}$.[6] An atomic sentence p is interpreted as a *test*, its meaning is a correctness check: $< a,b >\in \|p\|$ iff $a = b$. Negated sentences do not allow for establishing anaphoric links; in Groenendik's & Stokhof's terminology they are "externally static" wrt pronominal anaphora. Therefore, in *DPL*, negated sentences are also interpreted as a kind of test: $< a,b >\in \|\neg A\|$ iff $a = b$ and there is no c such that $< b,c >\in \|A\|$.

As far as the propositional level is concerned, the essentials of the dynamic interpretation in *DPL* are also available in the monoid semantics. We must define the dynamic interpretation of a formula A as a binary relation between information states: $\|A\| \subseteq I \times I$. Since we already have defined the partial order $\leq$ on I, we need not do this for each connective separately; one obvious and natural definition is:

$\|A\| = \{< a,b >| a \in v(A) \text{ and } a \leq b\}$.

According to this definition, the possible output states for an understanding of A in state a are the possible expansions (or prolongations) of a. This fits neatly into the above informational interpretation by means of monoid models. Distinguishing in the monoid semantics for our constructive minimal and constructive systems between positive valuations (truth conditions) and negative valuations (falsity conditions) even allows for a more fine-grained account which is sensitive to the difference between the verification and the falsification of a declarative sentence:

$\|A\|^+ = \{< a,b >| a \in v(A)^+ \text{ and } a \leq b\}$;
$\|A\|^- = \{< a,b >| a \in v(A)^- \text{ and } a \leq b\}$.

However, we can restrict ourselves to $\|A\|^+$, since $\|A\|^- = \| \sim A\|^+$.

[5]Groenendijk & Stokhof use '$\wedge$' instead of '$\circ$'.

[6]Van Benthem [1989] observes that the associative Lambek Calculus (with 'product' $\circ$) is sound wrt its natural relational interpretation. In this interpretation / and \ are defined as the following operations on binary relations:

$\|(A/B)\| = \{< a,b >| \forall c(< b,c >\in \|B\| \text{ implies } < a,c >\in \|A\|)\}$;
$\|(B \backslash A)\| = \{< a,b >| \forall c(< c,a >\in \|A\| \text{ implies } < c,b >\in \|B\|)\}$.

A sequent $X \rightarrow A$ is said to be valid according to this interpretation iff $\| \overset{\circ}{X} \| \subseteq \|A\|$ (note that X is not allowed to be the empty sequence). Došen [1990] points out that the associative Lambek Calculus, however, fails to be complete wrt the relational interpretation, if the operations /, \, and $\circ$ are defined on subsets of a set $\Gamma \times \Gamma$. Recently, Mikulás [1992] has proved completeness for subsets of a fixed transitive binary relation. In the same paper Mikulás and Andréka show that the associative Lambek Calculus with product is complete wrt the $\Gamma \times \Gamma$-type relational semantics, if either sequents with empty antecedents are admitted (and $\rightarrow A$ is said to be valid iff $\{< a,a >| a \in \Gamma\} \subseteq \|A\|$) or the following inference rules are added:

$A \rightarrow B \vdash C \rightarrow C \circ (A \backslash B)$, $A \rightarrow B \vdash C \rightarrow C \circ (B/A)$,
$A \rightarrow B \vdash C \rightarrow (A \backslash B) \circ C$, $A \rightarrow B \vdash C \rightarrow (B/A) \circ C$.

At this point one might object that, intuitively, the development of information states need not be *persistent*, i.e. preserve truth, or falsity, or both. True enough, expansion is not the only conceivable kind of development. The idea of a development of information states which fails to be truth or falsity preserving can also easily be realized in the monoid semantics. Define $\lhd \subseteq I \times I$ by $\{< a, b >| (\exists c \in I) a \cdot c = b\}$. We may then define the following non-persistent notion of development of information states:

$\|A\|^* = \{< a, b >| (\forall c \in v^+(A))\, a \lhd b\}$.[7]

Thus, dynamic semantics in the style of *DPL can* be done on *slomos* in interesting ways. Certainly, this is a topic to be further explored. For instance, given the informational structure of *slomos*, what are *natural dynamic operations*?

9.5 *Slomos* as an exhaustive format of abstract information structures

In this section we want to make use of the functional completeness results of Chapters 4 and 7. For this reason it will be instructive to discuss the philosophical significance of such theorems. Indubitably, the point of functional completeness results is that they establish *bounds*: classical negation $\neg$ and material implication $\supset$ are functionally complete for classical propositional logic *CPL* with its interpretation by two-valued truth-tables, and therefore in dealing with *CPL* we are entitled to *restrict* ourselves to $\neg$ and $\supset$. But, of course, we may also shift the perspective, start with, say, an axiomatization of the $\{\neg, \supset\}$-fragment of *CPL* and look for a semantics such that $\{\neg, \supset\}$ turns out as a functionally complete set of connectives. Clearly, these are two sides of the same coin, the difference lying in what we consider as basic and choose as our starting point. Accordingly, criticism of functional completeness theorems can be divided into (at least) two types. (a) One may complain about the *semantical theory* which is used. It may be too artificial, overly complicated, extremely poor, or whatever. (b) One may put into question the *selection of connectives* which are to be provided with a semantics that makes this collection emerge as functionally complete. The selection may be difficult to justify, or the justification may rest on controversial assumptions or convictions. An instantiation of criticism of type (a) is Kreisel's [1981] attitude towards the functional completeness results of McCullough [1971] and Zucker & Tragesser [1978]. Likewise our remarks concerning 'extraction' and 'infixation' as type forming operations of Categorial Grammar (see Chapter 4) can be used as a criticism of the semantical theory on which the functional completeness theorem for positive sequential propositional logic *PSPL* is based. Moreover, we have mentioned that the proof-theoretic interpretation of Chapters 4 and 7 does not cover operations like the **S4**-modality !. An example of avoiding this kind of criticsm by considering a certain set of connectives as fundamental is Schroeder-Heister's [1984] point of view that his functional completeness theorem for *IPL* should be considered as "demarcating the strength" (p. 1298) of intuitionistic conjunction, dis-

[7] In [Girard 1989] the idea can be found to conceive of the intensional, 'multiplicative' conjunction $\circ$ in linear logic as an update operation. Independently, similar ideas have been developed by Fuhrmann [1991]. Fuhrmann considers base systems which are considerably weaker than (certain fragments of) linear logic and associates additional inference rules with postulates of Gärdenfors' [1988] theory of belief revision.

junction, implication, and falsum. Similarly one could say that what McCullough [1971] does is specify the formats of truth-definitions for Kripke frames for IPL such that the set of intuitionistic connectives $\{\neg, \wedge, \vee, \supset\}$ resp. $\{\bot, \wedge, \vee, \supset\}$ turns out as functionally complete. No matter which position is adopted, the philosophical significance that is attached to a functional completeness theorem is thus always relative to either a set of connectives or a semantical theory. For our considerations we shall consider as basic not a certain set of connectives but rather the proof-theoretic semantics developed in Chapters 4 and 7, i.e. higher-level versions of Gentzen's sequent format with naturally obtained schemata for introducing connectives into premises and conclusions. We take it that this is a limitation to a respectable starting point.

Let us now, in order to have available a succinct notation, be explicit about the higher-level structural inference rules which we have added to the higher-level Gentzen caluli $\mathbf{G}^m$, $\mathbf{G}$, $\mathbf{GN}^-$, and $\mathbf{GN}$:

$$
\begin{array}{lll}
(\underline{\mathbf{P}}): & \mathsf{X}TU\mathsf{Y} \rightarrow T_1 \vdash \mathsf{X}UT\mathsf{Y} \rightarrow T_1, & T_1 \leftarrow \mathsf{X}TU\mathsf{Y} \vdash T_1 \leftarrow \mathsf{X}UT\mathsf{Y}; \\
(\underline{\mathbf{C}}): & \mathsf{X}TT\mathsf{Y} \rightarrow U \vdash \mathsf{X}T\mathsf{Y} \rightarrow U, & U \leftarrow \mathsf{X}TT\mathsf{Y} \vdash U \leftarrow \mathsf{X}T\mathsf{Y}; \\
(\underline{\mathbf{C}'}): & \mathsf{X}T\mathsf{Y}T\mathsf{Y}_1 \rightarrow U \vdash \mathsf{X}T\mathsf{Y}\mathsf{Y}_1 \rightarrow U, & U \leftarrow \mathsf{X}T\mathsf{Y}T\mathsf{Y}_1 \vdash U \leftarrow \mathsf{X}T\mathsf{Y}\mathsf{Y}_1, \\
 & \mathsf{X}T\mathsf{Y}T\mathsf{Y}_1 \rightarrow U \vdash \mathsf{X}\mathsf{Y}T\mathsf{Y}_1 \rightarrow U, & U \leftarrow \mathsf{X}T\mathsf{Y}T\mathsf{Y}_1 \vdash U \leftarrow \mathsf{X}\mathsf{Y}T\mathsf{Y}_1; \\
(\underline{\mathbf{E}}): & \mathsf{X}T\mathsf{Y} \rightarrow U \vdash \mathsf{X}TT\mathsf{Y} \rightarrow U, & U \leftarrow \mathsf{X}T\mathsf{Y} \vdash U \leftarrow \mathsf{X}TT\mathsf{Y}; \\
(\underline{\mathbf{E}'}): & \mathsf{X}T\mathsf{Y}\mathsf{Y}_1 \rightarrow U \vdash \mathsf{X}T\mathsf{Y}T\mathsf{Y}_1 \rightarrow U, & U \leftarrow \mathsf{X}T\mathsf{Y}\mathsf{Y}_1 \vdash U \leftarrow \mathsf{X}T\mathsf{Y}T\mathsf{Y}_1, \\
 & \mathsf{X}\mathsf{Y}T\mathsf{Y}_1 \rightarrow U \vdash \mathsf{X}T\mathsf{Y}T\mathsf{Y}_1 \rightarrow U, & U \leftarrow \mathsf{X}\mathsf{Y}T\mathsf{Y}_1 \vdash U \leftarrow \mathsf{X}T\mathsf{Y}T\mathsf{Y}_1; \\
(\underline{\mathbf{M}}): & \mathsf{X}\mathsf{Y} \rightarrow U \vdash \mathsf{X}T\mathsf{Y} \rightarrow U, & U \leftarrow \mathsf{X}\mathsf{Y} \vdash U \leftarrow \mathsf{X}T\mathsf{Y}.
\end{array}
$$

Let for $\Psi \in \{\mathbf{G}, \mathbf{G}^m, \mathbf{GN}^-, \mathbf{GN}\}$, $\Psi_{\underline{\Delta}}$ denote the result of extending Ψ by the higher-level structural inference rules in $\underline{\Delta} = \{\underline{R} \mid R \in \Delta,\ \Delta \subseteq \{\mathbf{P}, \mathbf{C}, \mathbf{C}', \mathbf{E}, \mathbf{E}', \mathbf{M}\}\}$. The following table summarizes earlier results:

$\mathbf{G}^m_{\underline{\Delta}} + (I) + (II)$	characterizes	$MSPL_{\Delta}$	is characterized by	$\mathsf{M}_{MSPL_{\Delta}}$
$\mathbf{G}_{\underline{\Delta}} + (I) + (II)$	characterizes	$ISPL_{\Delta}$	is characterized by	$\mathsf{M}_{ISPL_{\Delta}}$
$\mathbf{GN}^-_{\underline{\Delta}} + (I) - (IV)$	characterizes	$COSPL^-_{\Delta}$	is characterized by	$\mathsf{M}_{COSPL^-_{\Delta}}$
$\mathbf{GN}_{\underline{\Delta}} + (I) - (IV)$	characterizes	$COSPL_{\Delta}$	is characterized by	$\mathsf{M}_{COSPL_{\Delta}}$

Table 8.3: Characterization results.

In Table 8.3, in each case the proof-theoretic semantics mentioned is equivalent to the monoid semantics it is associated with. Hence, every connective F which is proof-theoretically definable is also explicitly definable in every monoid model of the appropriate model class M, since F is explicitly definable from the primitive connectives of the system characterized by M (and, moreover, since the monoid semantics is compositional). To this limited extent, i.e. insofar as we are content with propositional connectives which are proof-theoretically definable, monoid models resp. their underlying *slomos* form an exhaustive format of information models resp. abstract information structures.

We should, however, not forget about 'the other side of the coin'. Thus, there is an obvious

Problem Characterize the formats of truth-definitions for *slomos* with various properties such that $\{/, \backslash, \wedge, \circ, \vee, \top, \bot\}$ resp. $\{/, \backslash, \wedge, \circ, \vee, \top, \mathbf{t}, \bot\}$ resp. $\{\sim, /, \backslash, \wedge, \circ, /, \backslash, \mathbf{t}, \top\}$ turns out as a functionally complete set of connectives.

From a solution to this problem we can e.g. expect an answer to the interesting question whether in $COSPL^-$ there is a connective $F(A_1, \dots, A_n)$ definable from $\{\sim, /, \backslash, \wedge, \circ, \vee, \mathbf{t}, \top\}$ such that $v^-(F(A_1, \dots, A_n)) = \{a \mid 1 \not\leq a\}$.

Moreover, there are many *purely model-theoretic* notions of functional completeness over structures like *slomos*, see e.g. [van Benthem 1991].

9.6 Appendix: Applications

One use to which a sound and complete semantics usually is put, in particular, if it characterizes an undecidable logic, is the design of countermodels. By producing a countermodel we can e.g. show that $\not\vdash_{COSPL^-} \rightarrow\sim\sim (p_1 \backslash p_2) \rightleftarrows^+ \sim (p_1 \circ \sim p_2)$. The moniod model $< I, \cdot, \sqcap, 1, v_0^+, v_0^- >$ is (partially) specified as follows: $I = \{a, b\}$, $b \cdot b = a$, $a = 1$, $a \sqcap b = b$, $v_0^+(p_1) = \{a\}$, $v_0^+(p_2) = \{b\}$, and $v_0^-(p_1) = v_0^-(p_2) = I$. Then $v^+(\sim (p_1 \circ \sim p_2))$ $= v^+(\sim p_1 \circ p_2) = I$ whereas $v^+(\sim\sim (p_1 \backslash p_2)) = v^+(p_1 \backslash p_2) = \{b\}$. In the present appendix, we consider a few more applications of our monoid semantics; the first two of them make use of countermodels.

9.6.1 Monotonicity and paraconsistency

Recently, Urbas [1990] has suggested that genuine paraconsistent logics should lack the monotonicity rule **M** (alias weakening). A genuine paraconsistent logic should not only do without *ex contradictione sequitur quodlibet*, it should also fail to be 'explosive' wrt its connectives. In Johansson's MPL e.g. one can, by means of **M**, for every formula B prove $(A \wedge \neg A) \rightarrow \neg B$, i.e. although inconsistency does not lead to triviality wrt arbitrary formulas, it leads to triviality wrt arbitrary negated formulas. According to Urbas, the presence of **M** "renders it impossible to add anything but the most impoverished imitations of negation without thereby producing some version of explosiveness" [1990, p. 352]. In our *constructive* minimal systems, however, the presence of **M** does *not* imply explosiveness wrt one of the connectives.

Observation 9.22 There are monoid models for $COSPL^-_{\Delta'}$, $\mathbf{M} \in \Delta' \subseteq \{\mathbf{P}, \mathbf{C}, \mathbf{C}', \mathbf{E}, \mathbf{E}', \mathbf{M}\}$, and $L^\sim$-formulas A, B, C, such that

$$\begin{array}{ll} (A\wedge \sim A) \rightarrow\sim B, & (A\wedge \sim A) \rightarrow B \circ C, \\ (A\wedge \sim A) \rightarrow B \backslash C, & (A\wedge \sim A) \rightarrow C/B, \\ (A\wedge \sim A) \rightarrow B \wedge C, & (A\wedge \sim A) \rightarrow B \vee C, \end{array}$$

are not valid.

To be sure, Urbas' observations concerning the dependence of a system's explosiveness wrt certain connectives on the presence of **M** apply to systems with contraposition or they are true, if contraposition (as a rule) is assumed for the system resp. theories based on the system. Urbas shows that where in addition to contraposition as a rule "[w]eakening is present, all sentences in a theory are asserted to imply each other" [1990, p. 353]. He holds that attempting to avoid *ex contradictione* while retaining monotonicity "requires the abandonment also of symmetry-guaranteeing rules like [c]ontraposition, as well as desirable properties like the intersubstitutivity of provable equivalents. The result is logics which are paraconsistent only at the expense of important systemic properties and well-behaved connectives, especially negation. In short, *the cost of retaining [w]eakening ... is not worth it*" [1990, p. 353] (emphasis HW). From an informational perspective one couldn't disagree more. Both contraposition and intersubstitutivity of provable equivalents should be rejected, if positive and negative information are to be treated in their own right.[8] Indeed, intersubstitutivity of provable equivalents does not hold for $COSPL_{\Delta}^{-}$ and $COSPL_{\Delta}$ and moreover contraposition principles wrt $\sim$ fail to be valid in these logics. However, strong, constructive negation is a well-motivated and established concept. It can even be *discovered* in natural language, where it has e.g. made its way into the lexicon, as can be seen from pairs like **good** versus **bad** and **good** versus **not good**. Considering strong, constructive negation as a genuine negation thus reinforces the intuition that the notions of monotonicity and paraconsistency are independent of each other.

9.6.2 Negation in a formal system

At various places in the preceding chapters the concept of negation has been touched upon and discussed. In this subsection we shall look at negation from a somewhat more abstract perspective, viz. Gabbay's [1988] suggestion of a purely syntactical, proof-theoretic definition of what is negation in a formal system. We shall use the monoid semantics to show that negation as refutation in fact differs from negation as inconsistency, or, more concretely, that strong negation $\sim$ in the systems $COSPL_{\Delta}^{-}$ and $COSPL_{\Delta}$ provides a counterexample to Gabbay's characterization of negation as inconsistency.[9]

In terms of sequent calculus presentations, in [Gabbay 1988] formal systems are considered in which the sequent arrow $\rightarrow$ represents a syntactic consequence relation in the sense of Tarski and Scott, i.e., for all formulas A, B and finite sets of formulas Γ, Γ' we have:

(reflexivity)	$\vdash \Gamma \rightarrow A$ for $A \in \Gamma$;
(monotonicity)	if $\vdash \Gamma \rightarrow A$ and $\Gamma \subseteq \Gamma'$, then $\vdash \Gamma' \rightarrow A$;
(cut)	if $\vdash \Gamma \rightarrow A$ and $\vdash \Gamma \cup \{A\} \rightarrow B$, then $\vdash \Gamma \rightarrow B$.

Since on the lhs of $\rightarrow$ there appear *sets* of formulas, implicitly applications of the structural operations of permuting and contracting premise occurrences are permitted. This standard approach excludes substructural logics from the status of a formal system. We shall therefore require a sequent calculus to have among its rules merely the *logical* rules,

[8] This is not to say that one should dispense with contraposition in *every* application of strong negation. See e.g. [Nelson 1959] for systems with contraposible constructive negation.

[9] A more detailed investigation within the framework of structured logics can be found in [Gabbay & Wansing 1992].

(id) $\vdash A \to A$ and

(cut) $Y \to A \;\; XAZ \to B \vdash XYZ \to B$.[10]

To be precise, (id) and (cut) need not be *primitive* rules; if (id) resp. (cut) is not a primitive rule, then, however, it is required to be *admissible*, i.e. its addition must not increase the set of provable sequents. One may then add all kinds of structural inference rules, e.g. any combination taken from the collection $\Delta = \{\mathbf{P}, \mathbf{C}, \mathbf{C'}, \mathbf{E}, \mathbf{E'}, \mathbf{M}\}$ introduced in Chapter 2. Note that (id), (cut), and **M** are equivalent to (**reflexivity**), (**cut**), and (**monotonicity**) only in the presence of **P** and **C**.

Gabbay's [1988] central idea in defining a negation (as inconsistency) $*$ in a system is that $A \to *B$ is provable iff A and B *together* lead to some undesirable C from a set of 'unwanted' formulas $\theta*$. Now, let us assume *juxtaposition* as the operation for combining premise occurrences. The object-language counterpart of juxtaposition is intensional conjunction $\circ$. Let us therefore suppose that $\circ$ either is already in the language under consideration or that it can conservatively be added. Gabbay's basic definition which is appropriate e.g. for intuitionistic minimal, intuitionistic, and classical logic, can be reformulated in the more general framework as follows:

Definition 9.23 Let $\mathcal{L}$ be any formal system and let $*$ be a unary connective in the language L of $\mathcal{L}$. We say that $*$ is a negation in $\mathcal{L}$ iff there is a non-empty set of L-formulas $\theta*$ which is not the same as the set of all L-formulas, such that for every finite, possibly empty sequence X of L-formula occurrences and every L-formula A we have:

$$\vdash_{\mathcal{L}} X \to *A \text{ iff } (\exists B \in \theta*) (\vdash_{\mathcal{L}} XA \to B \text{ or } \vdash_{\mathcal{L}} AX \to B).$$[11]

If such a set of unwanted formulas $\theta*$ exists, it can always be taken as $\{C \mid \vdash_{\mathcal{L}} \to *C\}$, since by (id) the latter set is non-empty, if $*$ is a negation. In order to present a definiton which no longer refers to $\theta*$, Gabbay proves a lemma using (**monotonicity**). The presence of (**monotonicity**) is required, because (**cut**) is used instead of (cut).

Lemma 9.24 Let $*$ be a negation in $\mathcal{L}$ with $\theta*$ according to the definition of negation. Then for every finite, possibly empty sequence of L-formula occurrences X and every L-formula A, (a) and (b) resp. (a') and (b') are equivalent:

(a) $\vdash_{\mathcal{L}} XA \to C$ for some C such that $\vdash_{\mathcal{L}} \to *C$;

[10] Among the examples covered by the definition of negation in [Gabbay 1988] there nevertheless *is* a substructural logic, viz. an axiomatic presentation of $R_{\supset,\neg}$, i.e. relevant implicational logic $R_{\supset}$ together with the so-called Ackermann axioms for negation:

$$\begin{aligned} \mathrm{AN1}: &\quad (A \supset \neg B) \supset (B \supset \neg A), \\ \mathrm{AN2}: &\quad (A \supset \neg A) \supset \neg A, \\ \mathrm{AN3}: &\quad \neg\neg A \supset A. \end{aligned}$$

In order to treat this system as a formal system in the sense of Tarski and Scott, $\to$ has to be considered as representing relevant consequence in the sense of the relevant deduction theorem for $R_{\supset}$.

[11] Moreover, $\theta*$ should not contain theorems. Otherwise, for instance, with $\theta* = \{\top\}$ the unary operation $*A \stackrel{def}{=} (A \supset \top)$ would be a negation in IPL.

(b) $\vdash_{\mathcal{L}} XA \to B$ for some $B \in \theta*$;

(a') $\vdash_{\mathcal{L}} AX \to C$ for some C such that $\vdash_{\mathcal{L}} \to *C$;

(b') $\vdash_{\mathcal{L}} AX \to B$ for some $B \in \theta*$.

PROOF (b) implies (a), (b') implies (a'): Suppose that $B \in \theta*$. Since $\vdash_{\mathcal{L}} B \to B$, we have, by definition of $*$, $\vdash_{\mathcal{L}} \to *B$. (a) implies (b), (a') implies (b'): Suppose that for some C with $\vdash_{\mathcal{L}} \to *C$ we have $\vdash_{\mathcal{L}} XA \to C$ resp. $\vdash_{\mathcal{L}} AX \to C$. Since $\vdash_{\mathcal{L}} \to *C$, $\vdash_{\mathcal{L}} C \to B$ for some $B \in \theta*$. Applying (cut), one obtains $\vdash_{\mathcal{L}} XA \to B$ resp. $\vdash_{\mathcal{L}} AX \to B$. □

The definition of negation in a system can now be reformulated as follows:

Definition 9.25 A unary connective $*$ in the language L of a formal system $\mathcal{L}$ is said to be a negation iff for every finite, possibly empty sequence of L-formula occurrences X and every L-formula A the following holds:

$\vdash_{\mathcal{L}} X \to *A$ iff for some C s.t. $\vdash_{\mathcal{L}} \to *C$ we have $\vdash_{\mathcal{L}} XA \to C$ or $\vdash_{\mathcal{L}} AX \to C$.

It can easily be verifed that intuitionistic minimal, intuitionistic, and classical negation are negations in a system. In each case one may define $\theta* := \{\bot\}$. For negation in intuitionistic and classical propositional logic one may also define $\theta*$ as $\{\neg p \wedge p\}$, for some propositional variable p. According to our reformulated definition, $\neg^r$, $\neg^l$ are negations in $MSPL_\Delta$ and $ISPL_\Delta$: one may define $\theta\neg^r = \theta\neg^l := \{\bot\}$, or, in the case of $ISPL_\Delta$, $\theta\neg^r := \{\neg^r p \circ p\}$, $\theta\neg^l := \{p \circ \neg^l p\}$, for some propositional variable p.

Gabbay [1988] generalizes the basic definition of negation as inconsistency in the course of considering a number of formal systems with a recognized negation.[12] He also introduces one system which serves as a counterexample showing that the definition of negation remains non-trivial. The proof of non-triviality reveals a necessary condition for being a negation as inconsistency. In the present context it can be shown that, if $*$ is a negation in $\mathcal{L}$, then for each L_1-formula A, $\vdash_{\mathcal{L}_1} \to *(*A \circ A)$ or $\vdash_{\mathcal{L}_1} \to *(A \circ *A)$, where $\mathcal{L}_1$ is $\mathcal{L}$ conservatively extended by adding $\circ$:

	$\vdash_{\mathcal{L}_1} *A \to *A$		(id)
iff	$\vdash_{\mathcal{L}_1} *AA \to B$ or $\vdash_{\mathcal{L}_1} A*A \to B$	for some $B \in \theta*$	def. of neg.
iff	$\vdash_{\mathcal{L}_1} (*A \circ A) \to B$ or $\vdash_{\mathcal{L}_1} (A \circ *A) \to B$	for some $B \in \theta*$	$(\circ \to), (cut)$
iff	$\vdash_{\mathcal{L}_1} \to *(*A \circ A)$ or $\vdash_{\mathcal{L}_1} \to *(A \circ *A)$		def. of neg.,

for some set of L_1-formulas $\theta*$. Thus, if for some L-formula A there is a model for $\mathcal{L}_1$ that neither validates $*(*A \circ A)$ nor $*(A \circ *A)$, then $*$ cannot be a negation in $\mathcal{L}$. In other words, if $*$ is a negation in $\mathcal{L}$, then for each L_1-formula A, $*(*A \circ A)$ or $*(A \circ *A)$ is validated in every model for $\mathcal{L}_1$. Note that if one is dealing with partial valuations, then $*(*A \circ A)$ and $*(A \circ *A)$ need not be false (at a point) in a model in order not to be validated.

[12]The generalized definition refers to conservative extensions. It captures systems with $*$ which are not expressive enough to have a corresponding set $\theta*$. We have already remarked that, if $*$ is a negation in a formal system $\mathcal{L}$, then there are $\mathcal{L}$-theorems. By contraposition, $*$ cannot be a negation in $\mathcal{L}$, if $\mathcal{L}$ has no theorems. As has been pointed out by W. Rautenberg (personal communication), this would imply e.g. that classical negation $\neg$ is not a negation in the $\neg$-fragment of classical propositional logic. The definition in terms of conservative extensions takes care of cases like these (see also example E5 in [Gabbay 1988]).

Observation 9.26 Constructive negation $\sim$ fails to be a negation as inconsistency in $COPSL_{\Delta}^{-}$ and $COSPL_{\Delta}$.

Proof Take any *slomo* $\mathcal{F} = < \{1\}, \cdot, \cap, 1 >$, and define valuation functions v^+, v^- from the set of formulas into $2^{\{1\}}$, such that $\mathcal{M} = < \{1\}, \cdot, \cap, 1, v^+, v^- >$ is a monoid model for $COSPL$ and $v^+(p) = v^-(p) = \emptyset$. Then $v^-(\sim p \circ p) = v^-(p\circ \sim p) = \emptyset$. Therefore $1 \notin v^-(\sim p \circ p)$, $1 \notin v^-(p\circ \sim p)$, and hence $1 \notin v^+(\sim (\sim p \circ p))$, $1 \notin v^+(\sim (p\circ \sim p))$. Thus, neither $\sim (\sim p \circ p)$ nor $\sim (p\circ \sim p)$ is validated in $\mathcal{M}$. Since $\mathcal{M} \in \mathsf{M}_{COSPL_{\Delta}}$ and hence also $\mathcal{M} \in \mathsf{M}_{COSPL_{\Delta}^{-}}$, $\sim$ in $COSPL_{\Delta}$ and $COSPL_{\Delta}^{-}$ is not a negation as inconsistency. □

9.6.3 Beth's Theorem

A standard corollary to the interpolation theorem in classical and also intuitionistic logic is the definability theorem of Beth (see e.g. [Tennant 1978]), saying that certain notions of implicit and explicit definability coincide. We first shall consider 'extensional' notions of implicit and explicit definability, implicit e-definability resp. explicit e-definability. For these notions the proof of Beth's Theorem seems to depend on the presence of the traditional structural inference rules **P**, **C**, and **M**. We shall present a detailed proof of the theorem for $\Xi_{\{\mathbf{P},\mathbf{C},\mathbf{M}\}}$ ($\Xi \in \{MSPL, ISPL, COSPL^-, COSPL\}$) indicating each application of structural rules.

Let (X^p_q) denote the result of replacing every occurrence of p as a subformula in a formula from X by an occurrence of q, and let $var(X)$ denote the set of propositional variables that occur as subformulas in formulas from X.

Definition 9.27 A propositional variable $p \in var(X)$ is said to be implicitly e-definable in X wrt $\Xi_{\{\mathbf{P},\mathbf{C},\mathbf{M}\}}$ iff for every propositional variable $q \notin var(X)$ and every monoid model $< I, \cdot, \cap, 1, v_0 >$ resp. $< I, \cdot, \cap, 1, v_0^+, v_0^- >$ from $\mathsf{M}_{\Xi_{\{\mathbf{P},\mathbf{C},\mathbf{M}\}}}$:

$$v(X) \cap v((X^p_q)) \subseteq v(p \rightleftharpoons^+ q), \quad \text{if } \Xi \in \{MSPL, ISPL\}$$
$$\text{resp. } v^+(X) \cap v^+((X^p_q)) \subseteq v^+(p \rightleftharpoons q), \quad \text{if } \Xi \in \{COSPL^-, COSPL\}.$$

Definition 9.28 The variable p is said to be explicitly e-definable in X wrt $\Xi_{\{\mathbf{P},\mathbf{C},\mathbf{M}\}}$ iff there exists a formula A such that $var(A) \subseteq var(X) - \{p\}$ and:

$$\vdash_{\Xi_{\{\mathbf{P},\mathbf{C},\mathbf{M}\}}} X \rightarrow (p \rightleftharpoons^+ A), \quad \text{if } \Xi \in \{MSPL, ISPL\}$$
$$\text{resp. } \vdash_{\Xi_{\{\mathbf{P},\mathbf{C},\mathbf{M}\}}} X \rightarrow (p \rightleftharpoons A), \quad \text{if } \Xi \in \{COSPL^-, COSPL\}.$$

Theorem 9.29 A propositional variable p is explicitly e-definable in X wrt $\Xi_{\{\mathbf{P},\mathbf{C},\mathbf{M}\}}$ iff p is implicitly e-definable in X wrt $\Xi_{\{\mathbf{P},\mathbf{C},\mathbf{M}\}}$.

Proof Explicit implies implicit: Suppose p is explicitly e-definable in X wrt $\Xi_{\{\mathbf{P},\mathbf{C},\mathbf{M}\}}$. Then

$\vdash \overset{\circ}{X} \rightarrow (p \rightleftharpoons^+ A)$ resp. $\vdash \overset{\circ}{X} \rightarrow (p \rightleftharpoons A)$.

Consider the latter case. Clearly $\vdash (\overset{\circ}{X}{}^p_q) \rightarrow (q \rightleftharpoons A)$ and

$$\dfrac{\dfrac{\mathring{X}\to(p\rightleftharpoons A)}{\mathring{X}\wedge(\mathring{X}^p_q)\to(p\rightleftharpoons A)}\quad\dfrac{(\mathring{X}^p_q)\to(q\rightleftharpoons A)}{\mathring{X}\wedge(\mathring{X}^p_q)\to(q\rightleftharpoons A)}}{\mathring{X}\wedge(\mathring{X}^p_q)\to(p\rightleftharpoons A)\wedge(q\rightleftharpoons A).}$$

Using **C** this gives $\mathring{X}\wedge(\mathring{X}^p_q)\to(p\rightleftharpoons q)$. To see this, consider e.g. the following proof:

$$\dfrac{\dfrac{\dfrac{q/A\to q/A}{(A/p)\wedge(q/A)\to q/A}\quad\dfrac{A/p\to A/p}{(A/p)\wedge(q/A)\to A/p}}{(A/p)\wedge(q/A)\to(q/A)\wedge(A/p)}\quad\dfrac{\mathbf{C}\,\dfrac{\dfrac{\dfrac{q/A\to q/A}{(q/A)\wedge(A/p)\to q/A}\quad\dfrac{A/p\to A/p}{(q/A)\wedge(A/p)\to A/p}}{(q/A)\wedge(A/p)(q/A)\wedge(A/p)\to(q/A)\circ(A/p)}}{(q/A)\wedge(A/p)\to(q/A)\circ(A/p)}\quad\dfrac{\dfrac{\dfrac{p\to p\quad\dfrac{A\to A\quad q\to q}{(q/A)A\to q}}{(q/A)(A/p)p\to q}}{(q/A)(A/p)\to q/p}}{(q/A)\circ(A/p)\to q/p}}{(q/A)\wedge(A/p)\to q/p}}{(A/p)\wedge(q/A)\to q/p}$$

$$\vdots$$

$$\overline{(p\backslash A)\wedge(A\backslash p)\wedge(p/A)\wedge(A/p)\wedge(q\backslash A)\wedge(A\backslash q)\wedge(q/A)\wedge(A/q)\to q/p.}$$

By completeness, $\vdash\mathring{X}\wedge(\mathring{X}^p_q)\to(p\rightleftharpoons q)$ gives $v^+(X)\cap v^+(X^p_q)\subseteq v^+(p\rightleftharpoons q)$, for every monoid model $< I,\cdot,\cap,1,v_0^+,v_0^- >$ from $\mathsf{M}_{\Xi_{\{\mathbf{P},\mathbf{C},\mathbf{M}\}}}$. Implicit implies explicit: Assume that p is implicitly e-definable in X wrt $\Xi_{\{\mathbf{P},\mathbf{C},\mathbf{M}\}}$. Consider the constructive minimal and the constructive case. Then, by completeness, $\vdash\mathring{X}\wedge(\mathring{X}^p_q)\to(p\rightleftharpoons q)$, and therefore $\vdash\mathring{X}\,(\mathring{X}^p_q)\to(p\rightleftharpoons q)$:

$$\dfrac{\dfrac{\mathbf{M}\,\dfrac{\mathring{X}\to\mathring{X}}{\mathring{X}\mathring{X}^p_q\to\mathring{X}}\quad\mathbf{M}\,\dfrac{(\mathring{X}^p_q)\to(\mathring{X}^p_q)}{\mathring{X}\,(\mathring{X}^p_q)\to(\mathring{X}^p_q)}}{\mathring{X}\,(\mathring{X}^p_q)\to\mathring{X}\wedge(\mathring{X}^p_q)}\quad\mathring{X}\wedge(\mathring{X}^p_q)\to(p\rightleftharpoons q)}{\mathring{X}\,(\mathring{X}^p_q)\to(p\rightleftharpoons q).}$$

Now we can derive $p\,\mathring{X}\to q/(\mathring{X}^p_q)$ and $\sim p\,\mathring{X}\to\sim q/(\mathring{X}^p_q)$; consider e.g. the following derivation:

$$\underline{\mathring{X}(\mathring{X}^p_q)\to(p\backslash q)\wedge(q\backslash p)\wedge(p/q)\wedge(q/p)\wedge(\sim p\backslash\sim q)\wedge(\sim q\backslash\sim p)\wedge(\sim p/\sim p)\wedge(\sim q/\sim p)}$$

$$\vdots$$

$$\dfrac{\dfrac{\overline{\mathring{X}\,(\mathring{X}^p_q)\to p\backslash q}}{p\,\mathring{X}\,(\mathring{X}^p_q)\to q}}{p\,\mathring{X}\to q/(\mathring{X}^p_q).}$$

By the interpolation theorem there exists an interpolant A such that

(a) $\vdash p\,\mathring{X}\to A$

(b) $\vdash A\to q/(\mathring{X}^p_q)$

(c) $\vdash\sim p\,\mathring{X}\to A$

(d) $\vdash A \to\sim q/(\overset{\circ}{X}{}^{p}_{q})$ and

$var(A) \subseteq var(X) - \{p,q\}$. From (a) we obtain $\vdash X \to p\setminus A$ and, using **P**, $\vdash X \to A/p$. From (b) we obtain $\vdash (X^p_q) \to A \setminus q$ and, using **P**, $\vdash (X^p_q) \to q/A$ and hence also $\vdash X \to A\setminus p$ and $\vdash X \to p/A$. Similarly, (c) and (d) give $\vdash X \to\sim p\setminus A$, $\vdash X \to A/\sim p$, $\vdash X \to A\setminus\sim p$, and $\vdash X \to\sim p/A$. Altogether $\vdash X \to (p \rightleftharpoons A)$. □

With different, intensional notions of implicit and explicit definability (suggested, although not under these labels, by H. Schellinx), we can prove the Beth Theorem also in the absence of structural inference rules.

Definition 9.30 A propositional variable $p \in var(X)$ is said to be implicitly i-definable in X wrt Ξ_Δ iff for every propositional variable $q \notin var(X)$ and every monoid model $< I, \cdot, \cap, 1, v_0 >$ resp. $< I, \cdot, \cap, 1, v_0^+, v_0^- >$ from M_{Ξ_Δ}:

$$\begin{array}{ll} & v(X(X^p_q)) \subseteq v(p \rightleftharpoons q), \\ & v((X^p_q)X) \subseteq v(p \rightleftharpoons q), \quad \text{if } \Xi \in \{MSPL, ISPL\} \\ \text{resp.} & v^+(X(X^p_q)) \subseteq v^+(p \rightleftharpoons q), \\ & v^+((X^p_q)X) \subseteq v^+(p \rightleftharpoons q), \quad \text{if } \Xi \in \{COSPL^-, COSPL\}. \end{array}$$

Definition 9.31 The variable p is said to be explicitly i-definable in X wrt Ξ_Δ iff there exist formulas A, B such that $var(A) \subseteq var(X) - \{p\}$, $var(B) \subseteq var(X) - \{p\}$, and:

$$\begin{array}{ll} & \vdash_{\Xi_\Delta} X \to (p/A), \\ & \vdash_{\Xi_\Delta} X \to (A/p), \\ & \vdash_{\Xi_\Delta} X \to (p \setminus B), \\ & \vdash_{\Xi_\Delta} X \to (B \setminus p), \quad \text{if } \Xi \in \{MSPL, ISPL\} \\ \text{resp.} & \vdash_{\Xi_\Delta} X \to (p/A), \\ & \vdash_{\Xi_\Delta} X \to (A/p), \\ & \vdash_{\Xi_\Delta} X \to (p \setminus B), \\ & \vdash_{\Xi_\Delta} X \to (B \setminus p), \\ & \vdash_{\Xi_\Delta} X \to (\sim p/\sim A), \\ & \vdash_{\Xi_\Delta} X \to (\sim A/\sim p), \\ & \vdash_{\Xi_\Delta} X \to (\sim p\setminus\sim B), \\ & \vdash_{\Xi_\Delta} X \to (\sim B\setminus\sim p), \quad \text{if } \Xi \in \{COSPL^-, COSPL\}. \end{array}$$

Theorem 9.32 A propositional variable p is explicitly i-definable in X wrt Ξ_Δ iff p is implicitly i-definable in X wrt Ξ_Δ.

PROOF Explicit implies implicit: Suppose p is explicitly i-definable in X wrt $\Xi_{\{\mathbf{P},\mathbf{C},\mathbf{M}\}}$. This gives $\vdash (X^p_q)q \to A$ and $\vdash XA \to p$. By (cut), $\vdash X(X^p_q)q \to p$ and hence $\vdash X(X^p_q) \to p/q$. Similarly we get $\vdash X(X^p_q) \to q\setminus p$ and also $\vdash (X^p_q)X \to p \rightleftharpoons^+ q$, $\vdash X(X^p_q) \to p \rightleftharpoons q$, and $\vdash (X^p_q)X \to p \rightleftharpoons q$. Completeness gives $v(X(X^p_q)) \subseteq v(p \rightleftharpoons^+ q)$, $v((X^p_q)X) \subseteq v(p \rightleftharpoons^+ q)$ resp. $v^+(X(X^p_q)) \subseteq v^+(p \rightleftharpoons q)$, $v^+((X^p_q)X) \subseteq v^+(p \rightleftharpoons q)$. Implicit implies explicit: Consider again the constructive minimal and the constructive case. The only difference to the proof of the previous theorem is that completeness now directly gives

$$[*] \quad \vdash \overset{\circ}{X}\,(\overset{\circ}{X}{}^{p}_{q}) \rightarrow (p \rightleftharpoons q),$$
$$\vdash (\overset{\circ}{X}{}^{p}_{q})\,\overset{\circ}{X} \rightarrow (p \rightleftharpoons q).$$

Clearly, in the absence of **P** we cannot expect that [*] always leads to two interderivable interpolants. □

Bibliography

Ajdukiewicz, K., Die syntaktische Konnexität, *Studia Philosophica* 1 (1935), 1-27.

Akama, S., On the proof method for constructive falsity, *Zeitschrift für Mathematische Logik und Grundlagen der Mathematik* 34 (1988a), 385-392.

Akama, S., Constructive predicate logic with strong negation and model theory, *Notre Dame Journal of Formal Logic* 29 (1988b), 18-27.

Almukdad, A. & Nelson, D., Constructible falsity and inexact predicates, *Journal of Symbolic Logic* 49 (1984), 231-233.

Anderson A.R. & Belnap, N.D. Jr., *Entailment: The logic of relevance and necessity, Vol. I*, Princeton UP, Princeton, 1975.

Arruda, A.I., A survey of paraconsistent logic, in A.I. Arruda & R. Chuaqui, N.C.A. da Costa (eds.), *Mathematical Logic in Latin America*, North-Holland, Amsterdam, 1980.

Avron, A., The semantics and proof theory of linear logic, *Theoretical Computer Science* 57 (1988), 161-184.

Avron, A., Gentzenizing Schroeder-Heister's natural extension of natural deduction, *Notre Dame Journal of Formal Logic* 31 (1990), 127-135.

Barendregt, H.P., *The Lambda Calculus. Its Syntax and Semantics*, North-Holland, Amsterdam, 1984 (revised edition).

Belnap, N.D. Jr., A Useful Four-Valued Logic, in J.M. Dunn & G. Epstein (eds.), *Modern Uses of Multiple-Valued Logic*, Reidel, Dordrecht, 1977, 8-37.

Belnap, N.D. Jr., Display Logic, *Journal of Philosophical Logic* 11 (1982), 375-417.

van Benthem, J.F.A.K., *Essays in Logical Semantics*, Reidel, Dordrecht, 1986.

van Benthem, J.F.A.K., Semantic parallels in natural language and computation, in H.-D. Ebbinghaus et al. (eds.), *Logic Colloquium. Granada 1987*, North-Holland, Amsterdam, 1987.

van Benthem, J.F.A.K., The Lambek Calculus, in R. Oehrle et al. (eds.), *Categorial Grammars and Natural Language Structures*, Reidel, Dordrecht, 1988, 35-68.

van Benthem, J.F.A.K., *Language in Action*, North-Holland, Amsterdam, 1991.

Blamey, S., Partial Logic, in D.M. Gabbay & F. Guenthner (eds.), *Handbook of Philosophical Logic, Vol. III*, Reidel, Dordrecht, 1986, 1-70.

Buszkowski, W., The Logic of Types, in J. Srzednicki (ed.), *Initiatives in Logic*, Martinus Nijhoff, Dordrecht, 1987, 180-206.

Buszkowski, W., Generative power of Categorial Grammars, in R. Oehrle et al. (eds.), *Categorial Grammars and Natural Language Structures*, Reidel, Dordrecht, 1988, 69-94.

Buszkowski, W., Remarks on autoepistemic logic, typescript, 1989.

Church, A., The weak theory of implication, in A. Menne, A. Wilhelmy & H. Angsil (eds.), *Kontrolliertes Denken. Untersuchungen zum Logikkalkül und der Logik der Einzelwissenschaften*, Kommissions-Verlag Karl Alber, Munich, 1950, 22-37.

van Dalen, D., *Logic and Structure*, Springer Verlag, Berlin, 1983^2.

van Dalen, D., Intuitionistic Logic, in D.M. Gabbay & F. Guenthner (eds.), *Handbook of Philosophical Logic, Vol. III*, Reidel, Dordrecht, 1986, 225-339.

Došen, K., Sequent systems and groupoid models, I, *Studia Logica* 47 (1988), 353-389.

Došen, K., Sequent systems and groupoid models, II, *Studia Logica* 48 (1989), 41-65.

Došen, K., A brief survey of frames for the Lambek calculus, Report, Zentrum Philosophie und Wissenschaftstheorie 5-90, Universität Konstanz, 1990, to appear in *Zeitschrift für Mathematische Logik und Grundlagen der Mathematik*.

Dragalin, A., *Mathematical Intuitionism. Introduction to Proof Theory*, American Mathematical Society, Providence, 1988.

Dummett, M., *Elements of Intuitionism*, Oxford UP, Oxford, 1977.

Dunn, J.M., Relevance Logic and Entailment, in D.M. Gabbay & F. Guenthner (eds.), *Handbook of Philosophical Logic, Vol. III*, Reidel, Dordrecht, 1986, 177-224.

Fenstad J.-E., Halvorsen, P.-K., Langholm, T. & van Benthem, J.F.A.K., *Situations, Language and Logic*, Reidel, Dordrecht, 1987.

Fitch, F.B., *Symbolic Logic: an Introduction*, Ronald Press, New York, 1952.

Freudenthal, H., Zur intuitionistischen Deutung logischer Formeln, *Composito Mathematicae* 4 (1937), 112-116.

Friedman, H., Equality between functionals, in R. Parikh (ed.), *Logic Colloquium. Boston 1972-1973*, Springer, 1975, 22-37.

Fuhrmann, A., Cautious operations on premisses, Part I: Belief revision as substructural, typescript, 1991.

Gabbay, D.M., *Semantical Investigations in Heyting's Intuitionistic Logic*, Reidel, Dordrecht, 1981.

Gabbay, D.M., What is negation in a system?, in F.R. Drake & J.K. Truss (eds.), *Logic Colloquium '86*, Elsevier, Amsterdam, 1988, 95-112.

Gabbay, D.M., A general theory of structured consequence relations, Report, Imperial College of Science, Technology and Medicine, London, 1991, in: K. Došen & P. Schroeder-Heister (eds.), Proceedings of the conference on logics with restricted structural rules, Tübingen, October 1990, Oxford UP, to appear.

Gabbay, D.M. & de Queiroz, R.G.B., Extending the Curry-Howard interpretation to linear, relevant and other resource logics, *Journal of Symbolic Logic* 57 (1992), 1319-1365.

Gabbay, D.M. & Wansing, H., What is negation in a system? Part II: negation in structured consequence relations, 1992, in: A. Fuhrmann & H. Rott (eds.), Proceedings of the Konstanz Colloquium in Logic and Information, Konstanz, October 1992, De Gruyter, Berlin, to appear.

Gärdenfors, P., *Knowledge in Flux*, MIT Press, Cambridge, 1988.

Girard, J.-Y., Linear Logic, *Theoretical Computer Science* 50 (1987), 1-102.

Girard, J.-Y., Towards a geometry of interaction, in *Conference on Categories, Computer Science and Logic*, Contemporary Mathematics 92 (1989), 69-108.

Girard, J.-Y., Lafont, Y. & Taylor, P., *Proofs and Types*, CUP, Cambridge, 1989.

Gödel, K., Zum intuitionistischen Aussagenkalkül, *Anzeiger der Akademie der Wissenschaften in Wien. Math.-naturwiss. Klasse* 69 (1932), 42.

Groenendijk J. & Stokhof M., Dynamic Predicate Logic, *Linguistics & Philosophy*, 14 (1991), 39-101.

Grzegorczyk, A., A philosophically plausible formal interpretation of intuitionistic logic, *Indagationes Mathematicae* 26 (1964), 596-601.

Gurevich, Y., Intuitionistic logic with strong negation, *Studia Logica* 36 (1977), 49-59.

Heyting, A., *Intuitionism. An Introduction*, North-Holland, Amsterdam, 1956.

Hindley, J.R. & Seldin, J.P., *Introduction to Combinators and λ-Calculus*, Cambridge UP, Cambridge, 1986.

Horty, J., Thomason, R. & Touretzky, D., A skeptical theory of inheritance in nonmonotonic semantic nets, in *AAAI-87 (Proceedings of the Sixth National Conference on Artificial Intelligence)*, Vol. 2, Morgan Kaufman, Los Altos, 1987, 358-363.

Howard, W.A., The formulae-as-types notion of construction, in J.R. Hindley & J.P. Seldin (eds.), *To H.B. Curry: Essays on Combinatory Logic, Lambda Calculus and Formalism*, Academic Press, London, 1980, 479-490, (typescript, 1969).

Johansson, I., Der Minimalkalkül, ein reduzierter intuitionistischer Formalismus, *Composito Mathematicae* 4 (1937), 119-136.

Kanazawa, M., The Lambek Calculcus enriched with additional connectives, *Journal of Logic, Language and Information* 1 (1992), 141-171.

Kolmogorov, A.N., On the principle of the excluded middle (Russian), *Matematičeski Sbornik* 32 (1925), 646-667. English translation in J. van Heijenoort (ed.), *From Frege to Gödel. A source book in Mathematical Logic, 1879-1931*, Harvard UP, Cambridge Mass., 1967, 414-437.

Kreisel, G., Mathematical Logic, in T.L. Saaty (ed.), *Lectures on Modern Mathematics, Vol. III*, Wiley & Sons, New York, 1965, 95-195.

Kreisel, G., Constructivist approaches to logic, in E. Agazzi (ed.), *Modern Logic - A Survey*, Reidel, Dordrecht, 1981, 67-91.

Kripke, S.A., Semantical analysis of intuitionistic logic I, in J. Crossley & M. Dummett (eds.), *Formal Systems and Recursive Functions*, North-Holland, Amsterdam, 1965, 92-129.

von Kutschera, F., Die Vollständigkeit des Operatorensystems $\{\neg, \wedge, \vee, \supset\}$ für die intuitionistische Aussagenlogik im Rahmen der Gentzensemantik, *Archiv für Mathematische Logik und Grundlagenforschung* 11 (1968), 3-16.

von Kutschera, F., Ein verallgemeinerter Widerlegungsbegriff für Gentzenkalküle, *Archiv für Mathematische Logik und Grundlagenforschung* 12 (1969), 104-118.

Lambek, J., The mathematics of sentence structure, *American Mathematical Monthly* 65 (1958), 154-170.

Lambek, J., On the calculus of syntactic types, in R. Jakobson (ed.), *Structure of Language and its Mathematical Aspects*, American Mathematical Society, Providence, 1961, 166-178.

Lambek, J. & Scott, P., *Introduction to Higher Order Categorial Logic*, Cambridge UP, Cambridge, 1986.

Levesque, H.J., All I know: a study in autoepistemic logic, *Artificial Intelligence*, 42 (1990), 263-309.

López-Escobar, E.G.K., Refutability$'$ and elementary number theory, *Indagationes Mathematicae* 34 (1972), 362-374.

Lorenz, K., Dialogspiele als semantische Grundlage von Logikkalkülen, *Archiv für Mathematische Logik und Grundlagenforschung* 11 (1968), 32-55, 73-100.

Markov, A.A., Konstruktivnaja logika, *Usp. Mat. Nauk* 5 (1950), 187-188.

McCarthy, J., Circumscription - a form of non-monotonic reasoning, *Artificial Intelligence* 13 (1980), 27-39.

McCarty, C., Intuitionism: an introduction to a seminar. *Journal of Philosophical Logic* 12 (1983), 105-149.

McCollough, D.P., Logical connectives for intuitionistic propositional logic, *Journal of Symbolic Logic* 36 (1971), 15-20.

McKinsey, J.J.C., Proof of the independence of the primitive symbols of Heyting's calculus of propositions, *Journal of Symbolic Logic* 4 (1939), 155-158.

Mikulás, S., The completeness of the Lambek calculus with respect to relational semantics, ITLI Prebulication Series for Logic, Semantics and Philosophy of Language LP-92-03, University of Amsterdam, 1992.

Moh, S.-K., The deduction theorems and two new logical systems, *Methodos* 2 (1950), 56-75.

Moore, R.C., Semantical considerations on non-monotonic logic, *Artificial Intelligence* 25 (1985), 75-94.

Moortgat, M., *Categorial Investigations. Logical and Linguistic Aspects of the Lambek Calculus*, Foris, Dordrecht, 1988.

Moortgat, M., Discontinuous type constructors, Lecture held at the Second European Summer School in Language, Logic and Information, Leuven, August 1990.

Morrill, G., Grammar and logical types, in M. Stokhof and L. Torenvliet (eds.), *Proceedings of the Seventh Amsterdam Colloquium*, ITLI, Amsterdam, 1990, 429-450.

Nelson, D., Constructible falsity, *Journal of Symbolic Logic* 14 (1949), 16-26.

Nelson, D., Negation and separation of concepts in constructive systems, in A. Heyting (ed.), *Constructivity in Mathematics*. North-Holland, Amsterdam, 1959, 208-225.

Ono, H. & Komori, Y., Logics without the contraction rule, *Journal of Symbolic Logic* 50 (1985), 169-201.

Pearce, D., n reasons for choosing N, Report, Gruppe für Logik, Wissenstheorie und Information 14/91, Freie Universität Berlin, 1991.

Pearce, D. & Rautenberg, W., Propositional logic based on the dynamics of disbelief, in A. Fuhrmann & M. Morreau (eds.), *The Logic of Theory Change*, Springer, Berlin, 1991, 311-326.

Pearce, D. & Wagner, G., Reasoning with negative information I: strong negation in logic programs, in L. Haaparanta et al. (eds.), *Language, Knowledge, and Intentionality*, (*Acta Philosophica Fennica 49*), Helsinki, 1990, 430-453.

Pearce, D. & Wansing, H., On the methodology of possible worlds semantics, I: correspondence theory, *Notre Dame Journal of Formal Logic* 29 (1988), 482-496.

Popper, K.R., *Conjectures and Refutations: the Growth of Scientific Knowledge*, Routledge, London, 1963.

Pottinger, G., Normalization as a homomorphic image of cut-elimination, *Annals of Mathematical Logic* 12 (1977), 323-357.

Prawitz, D., Meaning and proofs: on the conflict between classical and intuitionistic logic, *Theoria* 43 (1977), 2-40.

Prawitz, D., Proofs and the meaning and completeness of the logical constants, in J. Hintikka et al. (ed.), *Essays on Mathematical and Philosophical Logic*, Reidel, Dordrecht, 1979, 25-40.

Rasiowa, H., *An Algebraic Approach to Non-classical Logics*, North-Holland, 1974.

Rautenberg, W., *Klassische und nichtklassische Aussagenlogik*, Vieweg, Braunschweig, 1979.

Reiter, R., A logic for default reasoning, *Artificial Intelligence* 13 (1980), 81-132.

Roorda, D., *Resource Logics*, PhD thesis, Institute for Logic, Language and Computation, University of Amsterdam, 1991.

Routley, R., Semantical analyses of propositional systems of Fitch and Nelson, *Studia Logica* 33 (1974), 283-298.

Ruitenburg, W., Constructive logic and the paradoxes, *Modern Logic* 1 (1991), 271-301.

Schroeder-Heister, P., A natural extension of natural deduction, *Journal of Symbolic Logic* 49 (1984), 1284-1300.

Schroeder-Heister, P., Structural Frameworks, Substructural Logics, and the Role of Elimination Inferences, in G. Huet & G. Plotkin (eds.), *Logical Frameworks*, Cambridge UP, Cambridge, 1991, 385-403.

Schütte, K., Der Interpolationssatz der intuitionistischen Prädikatenlogik, *Mathematische Annalen* 148 (1962), 192-200.

Slaney, J., Surendonk, T. & Girle R., Time, truth and logic, typescript, 1990.

Słupecki, J., Bryll, G. & Wybraniec-Skardowska, V., The theory of rejected propositions I, *Studia Logica* 29 (1971), 75-119.

Słupecki, J., Bryll, G. & Wybraniec-Skardowska, V., The theory of rejected propositions II, *Studia Logica* 30 (1972), 97-142.

Sundholm, G., Constructions, proofs and the meaning of the logical constants, *Journal of Philosophical Logic* 12 (1983), 151-172.

Tamura, S., The implicational fragment of R-mingle, *Proc. Japan Acad.* 47 (1971), 71-75.

Tanaka, K., A note on the proof method for constructive falsity, *Zeitschrift für Mathematische Logik und Grundlagen der Mathematik* 37 (1991), 63-64.

Tennant, N., *Natural Logic*, Edinburgh UP, Edinburgh, 1978.

Thijsse, E.G.C., Partial logic and modal logic: a systematic survey, Report, Institute for Language Technology and Artificial Intelligence, Tilburg University, 1990.

Thomason, R.H., A semantical study of constructive falsity, *Zeitschrift für Mathematische Logik und Grundlagen der Mathematik* 15 (1969), 247-257.

Troelstra A.S., Arend Heyting and his contribution to intuitionism, *Nieuw Archief voor Wiskunde* 24 (1981), 1-23.

Troelstra A.S., *Lectures on Linear Logic*, CSLI Lecture Notes No. 29, CSLI, Stanford, 1992.

Troelstra A.S. & van Dalen, D., *Constructivism in Mathematics, Vol. I*, North-Holland, Amsterdam, 1988.

Urbas, I., Paraconsistency, *Studies in Sovjet Thought* 39 (1990), 343-354.

Urquhart, A., Semantics for relevant logics, *Journal of Symbolic Logic* 37 (1972), 159-169.

Veltman, F., Data semantics, in J. Groenendijk & M. Stokhof (eds.), *Formal Methods in the Study of Language*, Mathematical Centre Tract 136, Amsterdam, 1981, 541-565.

de Vrijer, R., Strong normalization in $N-HA_p^\omega$, *Indagationes Mathematicae*, 49 (1987), 473-478.

Wagner, G., Logic programming with strong negation and inexact predicates, *Journal of Logic and Computation* 1 (1991), 835-859.

Wansing, H., Functional completeness for subsystems of intuitionistic propositional logic, Report, Gruppe für Logik, Wissenstheorie und Information 10/90, Freie Universität Berlin, 1990, revised version to appear in *Journal of Philosophical Logic* 22 (1993).

Wansing, H., Formulas-as-types for a hierarchy of sublogics of intuitionistic propositional logic, in D. Pearce & H. Wansing (eds.) *Nonclassical Logics and Information Processing*, Springer Lecture Notes in AI 619, Springer Verlag, Berlin, 1992, 125-145.

Zucker, J.I. & Tragesser R.S., The adequacy problem for inferential logic, *Journal of Philosophical Logic* 7 (1978), 501-516.

Summary

The present book considers from a logical point of view two central aspects of the concept of information, viz. information structure and information processing. Whereas information structure is regarded as a subject of model theory, information processing is treated as a matter of proof theory. In what follows the reader may find a brief, chapterwise summary.

Chapter 1 INTRODUCTION. This introductory chapter is mainly concerned with intuitionistic propositional logic *IPL*. Using a logical vocabulary that is suited for a systematic variation of structural inference rules, *IPL* is introduced as a sequent-calculus. Subsequently, two famous interpretations of *IPL* (and intuitionistic minimal propositional logic *MPL*) are reviewed: Kripke's and Grzegorczyk's. Both interpretations make use of models based on certain abstract information structures. The discussion of *IPL* is completed by a section on the so-called Brouwer-Heyting-Kolmogorov interpretation of the intuitionistic connectives. Moreover, the notion of a derivation in a sequent-calculus is explained in an appendix.

Chapter 2 GENERALIZATIONS. Taking up the presentation of *MPL* and *IPL* as logics of information structures, we discuss generalizations of these systems in two respects, viz. (i) in the direction of logical systems with a strong, constructive negation $\sim$, and (ii) in the direction of substructural logics, i.e. logics with a restricted set of structural inference rules. It is shown that both directions can be entered by a systematic criticism of the Brouwer-Heyting-Kolmogorov interpretation. The richer possibilities offered by the suggested generalizations are then illustrated by means of various examples. In an appendix, the previous considerations are used to suggest a few criteria for the notion of an informational interpretation of a given propositional logic.

Chapter 3 INTUITIONISTIC MINIMAL AND INTUITIONISTIC INFORMATION PROCESSING. We unfold a familiy of substructural subsystems of *MPL* and *IPL* and identify a number of known calculi within this family. In separat sections we investigate some important properties of the logical systems that have been introduced, viz. cut-eliminmation, decidability, and interpolation.

Chapter 4 FUNCTIONAL COMPLETENESS FOR SUBSTRUCTURAL SUBSYSTEMS OF *IPL*. Chapter 4 generalizes a certain proof-theoretic approach to the problem of functional completeness for a given propositional logic to the substructural case. We obtain functional completeness results for all subsystems of *MPL* and *IPL* under consideration wrt a proof theoretic semantics in terms of rule schemata for higher-level sequents. Finally, these results are applied to an operational extension of Categorial Grammar.

Chapter 5 FORMULAS-AS-TYPES FOR SUBSTRUCTURAL SUBSYSTEMS OF *IPL*. In this chapter, the so-called formulas-as-types notion of construction for intuitionistic implicational logic is extended to certain fragments of the substructural propositional logics that have been dealt with so far. Besides the encoding of proofs by typed terms and vice versa, the relation between cut-elimination and normalization of terms is considered. For some cases it is schown that these operations are homomorphic images of each other.

Chapter 6 CONSTRUCTIVE MINIMAL AND CONSTRUCTIVE INFORMATION PROCESSING. The point is reached at which the strong, constructive negation $\sim$ can be introduced into the intuitionistic minimal and intuitionistic systems. It is shown that in the presence of the ususal structural inference rules one in fact obtains David Nelson's propositional logics $\mathbf{N}^-$ and $\mathbf{N}$. Various peculiarities of constructive logic are discussed, in order to then take up again some earlier concerns: cut-elimination, decidability, interpolation, and the Brouwer-Heyting-Kolmogorov interpretation. The latter is extended from an interpretation of one particular system into a semantical framework for a broad spectrum of substructural propositional logics.

Chapter 7 FUNCTIONAL COMPLETENESS FOR SUBSTRUCTURAL SUBSYSTEMS OF $\mathbf{N}$. Chapter 7 is a 'constructive version' of Chapter 4. A corresponding proof theoretic approach towards functional completeness wrt $\mathbf{N}^-$ and $\mathbf{N}$ is generalized to substructural subsystems of these logics. The guiding idea is one of the earlier considerations on the Brouwer-Heyting-Kolmogorov interpretation: the notion of proof has to be supplemented by the notion of refutation (or disproof). There are some concluding remarks on negation in Categorial Grammar.

Chapter 8 THE CONSTRUCTIVE TYPED λ-CALCULUS λ^c AND FORMULAS-AS-TYPES FOR $\mathbf{N}^-$. Whereas Chapter 5 almost entirely dealt with syntactic aspects of the encoding of proofs by λ-terms, now a λ-term semantics for Nelson's $\mathbf{N}^-$ is developed and shown to be adequate. This result generalizes the usual λ-term semantics for intuitionistic implicational logic. Notably, strong negation $\sim$ is not captured by an operation on terms, but by means of a non-standard property of the term-calculus, viz. the non-unique typedness of terms.

Chapter 9 MONOID MODELS AND THE INFORMATIONAL INTERPRETATION OF SUBSTRUCTURAL PROPOSITIONAL LOGICS. This chapter links up with the two initial chapters. All the substructural propositional logics that have been introduced and investigated are now interpreted in so-called monoid models. The informational interpretation of these logics is based on the following reading of the structures $< I, \cdot, \sqcap, 1 >$ which underly the models: I is a set of information pieces (or states), $\cdot$ is the addition of information pieces, $\sqcap$ is the intersection of information pieces, and 1 is the initial, ideally the empty piece of information. In this way, the monoid semantics continues the well-known informational interpretations of models for *IPL* and related non-classical propositional logics. An extension by modal operators and the paradigm of dynamic interpretation are briefly considered. Eventually, the earlier functional completeness results are used, in order to explain that the monoid semantics in a certain sense constitutes an exhaustive format of abstract information structures. An appendix is devoted to a few applications of monoid models, among others Beth's definability theorem and the proof that $\sim$ fails to be a negation as inconsistency.

Index

The following index is not in every respect systematic or even 'complete'. However, it is intended to be user-friendly.

Printing: Weihert-Druck GmbH, Darmstadt
Binding: Buchbinderei Schäffer, Grünstadt

Lecture Notes in Artificial Intelligence (LNAI)

Vol. 567: H. Boley, M. M. Richter (Eds.), Processing Declarative Kowledge. Proceedings, 1991. XII, 427 pages. 1991.

Vol. 568: H.-J. Bürckert, A Resolution Principle for a Logic with Restricted Quantifiers. X, 116 pages. 1991.

Vol. 587: R. Dale, E. Hovy, D. Rösner, O. Stock (Eds.), Aspects of Automated Natural Language Generation. Proceedings, 1992. VIII, 311 pages. 1992.

Vol. 590: B. Fronhöfer, G. Wrightson (Eds.), Parallelization in Inference Systems. Proceedings, 1990. VIII, 372 pages. 1992.

Vol. 592: A. Voronkov (Ed.), Logic Programming. Proceedings, 1991. IX, 514 pages. 1992.

Vol. 596: L.-H. Eriksson, L. Hallnäs, P. Schroeder-Heister (Eds.), Extensions of Logic Programming. Proceedings, 1991. VII, 369 pages. 1992.

Vol. 597: H. W. Guesgen, J. Hertzberg, A Perspective of Constraint-Based Reasoning. VIII, 123 pages. 1992.

Vol. 599: Th. Wetter, K.-D. Althoff, J. Boose, B. R. Gaines, M. Linster, F. Schmalhofer (Eds.), Current Developments in Knowledge Acquisition - EKAW '92. Proceedings. XIII, 444 pages. 1992.

Vol. 604: F. Belli, F. J. Radermacher (Eds.), Industrial and Engineering Applications of Artificial Intelligence and Expert Systems. Proceedings, 1992. XV, 702 pages. 1992.

Vol. 607: D. Kapur (Ed.), Automated Deduction – CADE-11. Proceedings, 1992. XV, 793 pages. 1992.

Vol. 610: F. von Martial, Coordinating Plans of Autonomous Agents. XII, 246 pages. 1992.

Vol. 611: M. P. Papazoglou, J. Zeleznikow (Eds.), The Next Generation of Information Systems: From Data to Knowledge. VIII, 310 pages. 1992.

Vol. 617: V. Mařík, O. Štěpánková, R. Trappl (Eds.), Advanced Topics in Artificial Intelligence. Proceedings, 1992. IX, 484 pages. 1992.

Vol. 619: D. Pearce, H. Wansing (Eds.), Nonclassical Logics and Information Processing. Proceedings, 1990. VII, 171 pages. 1992.

Vol. 622: F. Schmalhofer, G. Strube, Th. Wetter (Eds.), Contemporary Knowledge Engineering and Cognition. Proceedings, 1991. XII, 258 pages. 1992.

Vol. 624: A. Voronkov (Ed.), Logic Programming and Automated Reasoning. Proceedings, 1992. XIV, 509 pages. 1992.

Vol. 627: J. Pustejovsky, S. Bergler (Eds.), Lexical Semantics and Knowledge Representation. Proceedings, 1991. XII, 381 pages. 1992.

Vol. 633: D. Pearce, G. Wagner (Eds.), Logics in AI. Proceedings. VIII, 410 pages. 1992.

Vol. 636: G. Comyn, N. E. Fuchs, M. J. Ratcliffe (Eds.), Logic Programming in Action. Proceedings, 1992. X, 324 pages. 1992.

Vol. 638: A. F. Rocha, Neural Nets. A Theory for Brains and Machines. XV, 393 pages. 1992.

Vol. 642: K. P. Jantke (Ed.), Analogical and Inductive Inference. Proceedings, 1992. VIII, 319 pages. 1992.

Vol. 659: G. Brewka, K. P. Jantke, P. H. Schmitt (Eds.), Nonmonotonic and Inductive Logic. Proceedings, 1991. VIII, 332 pages. 1993.

Vol. 660: E. Lamma, P. Mello (Eds.), Extensions of Logic Programming. Proceedings, 1992. VIII, 417 pages. 1993.

Vol. 667: P. B. Brazdil (Ed.), Machine Learning: ECML – 93. Proceedings, 1993. XII, 471 pages. 1993.

Vol. 671: H. J. Ohlbach (Ed.), GWAI-92: Advances in Artificial Intelligence. Proceedings, 1992. XI, 397 pages. 1993.

Vol. 679: C. Fermüller, A. Leitsch, T. Tammet, N. Zamov, Resolution Methods for the Decision Problem. VIII, 205 pages. 1993.

Vol. 681: H. Wansing, The Logic of Information Structures. IX, 163 pages. 1993.

Vol. 689: J. Komorowski, Z. W. Raś (Eds.), Methodologies for Intelligent Systems. Proceedings, 1993. XI, 653 pages. 1993.

Vol. 695: E. P. Klement, W. Slany (Eds.), Fuzzy Logic in Artificial Intelligence. Proceedings, 1993. VIII, 192 pages. 1993.

Vol. 698: A. Voronkov (Ed.), Logic Programming and Automated Reasoning. Proceedings, 1993. XIII, 386 pages. 1993.

Vol. 699: G.W. Mineau, B. Moulin, J.F. Sowa (Eds.), Conceptual Graphs for Knowledge Representation. Proceedings, 1993. IX, 451 pages. 1993.

Lecture Notes in Computer Science

Vol. 661: S. J. Hanson, W. Remmele, R. L. Rivest (Eds.), Machine Learning: From Theory to Applications. VIII, 271 pages. 1993.

Vol. 662: M. Nitzberg, D. Mumford, T. Shiota, Filtering, Segmentation and Depth. VIII, 143 pages. 1993.

Vol. 663: G. v. Bochmann, D. K. Probst (Eds.), Computer Aided Verification. Proceedings, 1992. IX, 422 pages. 1993.

Vol. 664: M. Bezem, J. F. Groote (Eds.), Typed Lambda Calculi and Applications. Proceedings, 1993. VIII, 433 pages. 1993.

Vol. 665: P. Enjalbert, A. Finkel, K. W. Wagner (Eds.), STACS 93. Proceedings, 1993. XIV, 724 pages. 1993.

Vol. 666: J. W. de Bakker, W.-P. de Roever, G. Rozenberg (Eds.), Semantics: Foundations and Applications. Proceedings, 1992. VIII, 659 pages. 1993.

Vol. 667: P. B. Brazdil (Ed.), Machine Learning: ECML – 93. Proceedings, 1993. XII, 471 pages. 1993. (Subseries LNAI).

Vol. 668: M.-C. Gaudel, J.-P. Jouannaud (Eds.), TAPSOFT '93: Theory and Practice of Software Development. Proceedings, 1993. XII, 762 pages. 1993.

Vol. 669: R. S. Bird, C. C. Morgan, J. C. P. Woodcock (Eds.), Mathematics of Program Construction. Proceedings, 1992. VIII, 378 pages. 1993.

Vol. 670: J. C. P. Woodcock, P. G. Larsen (Eds.), FME '93: Industrial-Strength Formal Methods. Proceedings, 1993. XI, 689 pages. 1993.

Vol. 671: H. J. Ohlbach (Ed.), GWAI-92: Advances in Artificial Intelligence. Proceedings, 1992. XI, 397 pages. 1993. (Subseries LNAI).

Vol. 672: A. Barak, S. Guday, R. G. Wheeler, The MOSIX Distributed Operating System X, 221 pages. 1993.

Vol. 673: G. Cohen, T. Mora, O. Moreno (Eds.), Applied Algebra, Algebraic Algorithms and Error-Correcting Codes. Proceedings, 1993. X, 355 pages 1993.

Vol. 674: G. Rozenberg (Ed.), Advances in Petri Nets 1993. VII, 457 pages. 1993.

Vol. 675: A. Mulkers, Live Data Structures in Logic Programs. VIII, 220 pages. 1993.

Vol. 676: Th. H. Reiss, Recognizing Planar Objects Using Invariant Image Features. X, 180 pages. 1993.

Vol. 677: H. Abdulrab, J.-P. Pécuchet (Eds.), Word Equations and Related Topics, Proceedings, 1991. VII, 214 pages. 1993.

Vol. 678: F. Meyer auf der Heide, B. Monien, A. L. Rosenberg (Eds.), Parallel Architectures and Their Efficient Use. Proceedings, 1992. XII, 227 pages. 1993.

Vol. 679: C. Fermüller, A. Leitsch, T. Tammet, N. Zamov, Resolution Methods for the Decision Problem. VIII, 205 pages. 1993. (Subseries LNAI).

Vol. 681: H. Wansing, The Logic of Information Structures. IX, 163 pages. 1993. (Subseries LNAI).

Vol. 682: B. Bouchon-Meunier, L. Valverde, R. R. Yager (Eds.), IPMU '92 – Advanced Methods in Artificial Intelligence. Proceedings, 1992. IX, 367 pages. 1993.

Vol. 683: G.J. Milne, L. Pierre (Eds.), Correct Hardware Design and Verification Methods. Proceedings, 1993. VIII, 270 Pages. 1993.

Vol. 684: A. Apostolico, M. Crochemore, Z. Galil, U. Manber (Eds.), Combinatorial Pattern Matching. Proceedings, 1993. VIII, 265 pages. 1993.

Vol. 685: C. Rolland, F. Bodart, C. Cauvet (Eds.), Advanced Information Systems Engineering. Proceedings, 1993. XI, 650 pages. 1993.

Vol. 686: J. Mira, J. Cabestany, A. Prieto (Eds.), New Trends in Neural Computation. Procedings, 1993. XVII, 746 pages. 1993.

Vol. 687: H. H. Barrett, A. F. Gmitro (Eds.), Information Processing in Medical Imaging. Proceedings, 1993. XVI, 567 pages. 1993.

Vol. 688: M. Gauthier (Ed.), Ada - Europe '93. Proceedings, 1993. VIII, 353 pages. 1993.

Vol. 689: J. Komorowski, Z. W. Ras (Eds.), Methodologies for Intelligent Systems. Proceedings, 1993. XI, 653 pages. 1993. (Subseries LNAI).

Vol. 690: C. Kirchner (Ed.), Rewriting Techniques and Applications. Proceedings, 1993. XI, 488 pages. 1993.

Vol. 691: M. Ajmone Marsan (Ed.), Application and Theory of Petri Nets 1993. Proceedings, 1993. IX, 591 pages. 1993.

Vol. 692: D. Abel, B.C. Ooi (Eds.), Advances in Spatial Databases. Proceedings, 1993. XIII, 529 pages. 1993.

Vol. 693: P. E. Lauer (Ed.), Functional Programming, Concurrency, Simulation and Automated Reasoning. Proceedings, 1991/1992. XI, 398 pages. 1993.

Vol. 694: A. Bode, M. Reeve, G. Wolf (Eds.), PARLE '93. Parallel Architectures and Languages Europe. Proceedings, 1993. XVII, 770 pages. 1993.

Vol. 695: E. P. Klement, W. Slany (Eds.), Fuzzy Logic in Artificial Intelligence. Proceedings, 1993. VIII, 192 pages. 1993. (Subseries LNAI).

Vol. 697: C. Courcoubetis (Ed.), Computer Aided Verification. Proceedings, 1993. IX, 504 pages. 1993.

Vol. 698: A. Voronkov (Ed.), Logic Programming and Automated Reasoning. Proceedings, 1993. XIII, 386 pages. 1993. (Subseries LNAI).